Books are to be returned on or before
the last date below.

29 JAN 1996   11 MAR 1998

New Methods of Polymer Synthesis

Edited by

J.R. EBDON
Polymer Centre
University of Lancaster

Blackie
Glasgow and London

Published in the USA by
Chapman and Hall
New York

Blackie & Son Ltd.
Bishopbriggs, Glasgow G64 2NZ
and
7 Leicester Place, London WC2H 7BP

Published in the USA by
Chapman and Hall
a division of Routledge, Chapman and Hall, Inc.
29 West 35th Street, New York, NY 10001-2291

© 1991 Blackie & Son Ltd.
First published 1991

All rights reserved.
No part of this publication may be reproduced,
stored in a retrieval system, or transmitted,
in any form or by any means—graphic,
electronic or mechanical, including photocopying,
recording, taping—without the
written permission of the publishers

British Library Cataloguing in Publication Data

New methods of polymer synthesis.
1. Polymerisation
I. Ebdon, J.R.
547.28

ISBN 0-216-92974-1

Library of Congress Cataloging-in-Publication Data

New methods of polymer synthesis / edited by J.R. Ebdon.
  p.   cm.
Includes bibliographical references and index.
ISBN 0-412-02471-3
1. Polymerization.   I. Ebdon, J.R.
QD281.P6N48 1991
668.9—dc20   90-43657
                                                          CIP

ANDERSONIAN LIBRARY

1 3. JUN 91

UNIVERSITY OF STRATHCLYDE

Typesetting by Thomson Press (India) Limited, New Delhi
Printed in Great Britain by Thomson Litho Ltd, East Kilbride, Scotland

Preface

Most practitioners and students of polymer chemistry are familiar, in general terms at least, with the established methods of polymer synthesis – radical, anionic, cationic and coordination addition polymerization, and stepwise condensation and rearrangement polymerization. These methods are used to synthesize the majority of polymers used in the manufacture of commercially important plastics, fibres, resins and rubbers, and are covered in most introductory polymer chemistry textbooks and in most undergraduate and graduate courses on polymer science.

Fewer polymer chemists, however, have much familiarity with more recent developments in methods of polymer synthesis, unless they have been specifically involved for some time in the synthesis of speciality polymers. These developments include not only refinements to established methods but also new mechanisms of polymerization, such as group transfer and metathesis polymerization and novel non-polymerization routes to speciality polymers involving, for example, the chemical modification of preformed polymers or the linking together of short terminally functionalized blocks.

Until now, those wishing to obtain an overview of recent developments in methods of polymer synthesis have had to consult a large number and wide variety of sources, including primary scientific journals, conference reports and specialist review articles. This book attempts to make the process of obtaining such an overview easier, by bringing together up-to-date reviews of a number of key areas. Inevitably, not all developments in methods of polymer synthesis of interest and importance can be included in a single book, if it is to be of manageable proportions. However, the topics have been chosen with care to cover a wide range of chemistry and, hopefully, therefore to serve a variety of needs. In particular, it is hoped that the book will be of use to academic and industrial researchers anxious to keep abreast of recent developments in polymer science, to organic chemists and others drawn into the synthesis of novel polymers, to undergraduate and graduate students seeking background information on newer methods of polymer synthesis, and last, but by no means least, to university teachers anxious to update their teaching of polymer synthesis, especially at the graduate level. Indeed it was my attempts to teach aspects of advanced polymer synthesis to graduate students at the University of Lancaster that first persuaded me that there was a need for a book which brought together in one place information on a variety of recent work.

No book covering such a range of topics can be written with authority by one person, and I am grateful to my co-authors for their expert contributions, without which this book could not have been brought to fruition. I am grateful also to the publishers for their help and guidance with the project, to my colleagues within the Polymer Centre at Lancaster for their encouragement and, particularly, to my family for their patient and cheerful moral support throughout.

J.E.

Contents

1 Developments in polymerization — 1
J.R. EBDON

1.1 Introduction — 1
1.2 Radical addition polymerization — 2
    1.2.1 General features — 2
    1.2.2 Iniferters — 3
    1.2.3 Template polymerization — 5
    1.2.4 Ring-opening polymerization — 6
    1.2.5 Copolymerization — 8
1.3 Ionic and coordination addition polymerization — 9
    1.3.1 General background and developments — 9
    1.3.2 Living anionic homo- and copolymerization of oxiranes — 12
    1.3.3 The activated monomer mechanism of oxirane polymerization — 13
    1.3.4 Living carbocationic polymerization of vinyl ethers — 14
1.4 Stepwise polymerization — 14
    1.4.1 General background — 14
    1.4.2 Aromatic polyesters and polyamides — 15
    1.4.3 Polysulphones and polyketones — 16
    1.4.4 Starburst polymers: dendrimers and arborols — 17
    1.4.5 Biosynthetic routes to polymers — 18
References — 19

2 Group transfer polymerizations — 22
G.C. EASTMOND and O.W. WEBSTER

2.1 Introduction — 22
    2.1.1 Background — 22
    2.1.2 Terminology — 24
    2.1.3 Nucleophilic addition — 25
    2.1.4 Silyl ketene acetals — 27
2.2 Features of group transfer polymerization — 28
    2.2.1 General features — 28
    2.2.2 Reaction conditions — 30
    2.2.3 Initiators — 31
    2.2.4 Catalysts — 35
    2.2.5 Monomers — 36
    2.2.6 Termination — 39
    2.2.7 Polymer microstructures — 42
2.3 Kinetics and mechanism — 43
    2.3.1 Polymerization of methacrylates — 43
    2.3.2 Polymerization of acrylates — 55
2.4 Aldol group transfer polymerization — 57
2.5 Copolymers — 59
    2.5.1 Random copolymers — 60
    2.5.2 Block copolymers — 61

	2.5.3	Coupling	64
	2.5.4	Other routes	65
2.6	Telechelics		66
2.7	Related and anionic polymerizations		67
2.8	Applications		70
References			72

3 Ring-opening metathesis polymerization of cyclic alkenes   76
A.J. AMASS

3.1	Scope of ring-opening metathesis polymerization		76
3.2	Monomers for ring-opening metathesis polymerization		77
3.3	Catalysts for the ring-opening polymerization of cycloalkenes		80
3.4	Mechanism of ring-opening metathesis polymerization		84
	3.4.1	Site of olefin metathesis	84
	3.4.2	Pairwise mechanisms	86
	3.4.3	Non-pairwise mechanisms	87
	3.4.4	Metallocyclobutanes	91
	3.4.5	Transition metal carbenes	92
3.5	Molecular weight distribution in polyalkenylenes		94
3.6	Stereochemistry of ring-opening metathesis polymerization		97
	3.6.1	Stereoisomerism of polyalkenylenes	98
	3.6.2	Stereoisomerism in norbornene polymers	98
	3.6.3	Substituted norbornenes	99
3.7	Thermodynamics of ring-opening polymerization		104
References			105

4 Transformation reactions   107
M.J. STEWART

4.1	Introduction		107
4.2	Historical development of transformation reactions		110
4.3	Transformations between anionic and cationic polymerization		112
	4.3.1	Anion to cation transformations	112
	4.3.2	Cation to anion transformations	116
	4.3.3	Anion to cation coupling reactions	118
4.4	Transformations between ionic and free radical polymerization		120
	4.4.1	Anion to radical transformations	120
	4.4.2	Cation to radical transformations	126
	4.4.3	Radical to ionic transformations	128
4.5	Transformations involving other modes of polymerization		129
	4.5.1	Anion to Ziegler–Natta transformations	129
	4.5.2	Anion to metathesis transformations	130
	4.5.3	Ziegler–Natta to radical transformations	131
	4.5.4	Group transfer to radical transformations	132
	4.5.5	Metathesis to aldol-group transfer transformations	132
	4.5.6	Radical to active monomer transformations	133
	4.5.7	Transformations between radical polymerization and polypeptide synthesis	134
4.6	Conclusions		134
References			135

5 Chemical modification of preformed polymers   138
F.G. THORPE

5.1	Introduction		138
5.2	Chemical reactions of polymers: general aspects		139
5.3	Modification of polymers		140
	5.3.1	Modifications on the polymer main chain	140

	5.3.2	Modifications of pendant groups	145
	5.3.3	Miscellaneous systems	157
	5.3.4	Concluding remarks	159
References			159

6 Terminally reactive oligomers: telechelic oligomers and macromers   162
J.R. EBDON

6.1	Introduction		162
6.2	The synthesis of terminally reactive oligomers		164
	6.2.1	By anionic polymerization	165
	6.2.2	By group transfer polymerization	168
	6.2.3	By cationic polymerization	169
	6.2.4	By radical polymerization	174
	6.2.5	By stepwise polymerization	178
	6.2.6	By controlled polymer chain scission	181
	6.2.7	Modifications of end-groups of telechelic oligomers	183
6.3	Reactions and reactivity of telechelic oligomers and macromers		185
	6.3.1	Telechelic oligomers	185
	6.3.2	Macromers	187
6.4	Uses of terminally functionalized oligomers		188
	6.4.1	Telechelic oligomers	188
	6.4.2	Macromers	190
References			191

Index   197

Contributors

A.J. Amass Department of Chemical Engineering and Applied Chemistry, Aston University, Birmingham B4 7ET, UK

G.C. Eastmond Department of Chemistry, University of Liverpool, P.O. Box 147, Liverpool L69 3BX, UK

J.R. Ebdon The Polymer Centre, Chemistry Building, Lancaster University, Lancaster LA1 4YA, UK

M.J. Stewart Royal Armament Research and Development Establishment, Powdermill Lane, Waltham Abbey, Essex EN9 1AX, UK

F.G. Thorpe The Polymer Centre, Chemistry Building, Lancaster University, Lancaster LA1 4YA, UK

O.W. Webster Central Research and Development Department, E.I. DuPont de Nemours and Co. Inc., Experimental Station, Wilmington, Delaware 19880, USA

1 Developments in polymerization

J.R. EBDON

1.1 Introduction

The foundations for the understanding of the major mechanisms of polymer synthesis, that is, radical, anionic, cationic and coordination addition polymerizations, and stepwise condensation and rearrangement polymerizations, were laid during the period from 1930 to 1960. Of the mechanisms mentioned above, which we can regard as being the primary methods for the synthesis of polymers, the most recently discovered is coordination (or Ziegler–Natta) polymerization.[1] Today, literally hundreds of different types of polymers are produced by these mechanisms, ranging from bulk thermoplastics such as polyethylene, polystyrene and poly(vinyl chloride), through fibre-forming materials such as polyacrylonitrile, polyesters and polyamides (nylons), elastomers such as *cis*-1,4-polyisoprene, polychloroprene and ethylene-propylene copolymers, to high-performance engineering materials such as polycarbonates, epoxy resins, polyimides and aromatic polyethers and polysulphones. Much of the research and development in polymer synthesis over the past thirty years has been concerned with improvements to, and refinements of, the above primary methods aimed at achieving one or more of the following objectives:

1. the better control of polymer molecular weight and molecular weight distribution;
2. the better control of polymer microstructure, i.e. stereoisomerism, geometrical isomerism, regioisomerism and, in the case of copolymers, also composition and monomer sequence distribution; and
3. the generation of polymers based on new monomers and hence of polymers with new properties and new applications.

It is not possible in a single chapter, or even in a single book, to review adequately all of the significant advances in the understanding and application of the primary methods of polymerization. Instead, this chapter singles out for consideration just a few topics of recent and current interest. The topics chosen are inevitably a very personal, perhaps even idiosyncratic, selection and apologies are offered to those whose important work might appear to have been overlooked. No apologies are offered, however, for devoting most space to discussion of aspects of radical addition and stepwise polymerizations; these particular mechanisms are still the most important in view of their wide utility. In discussing these topics, emphasis is placed upon the structures of the polymers formed and how these are controlled by the mechanism of polymerization, as in objectives (1), (2) and (3) above.

For more thorough coverage of these and other topics, readers are referred to recently published encyclopaedic works and the references cited therein.[2,3]

1.2 Radical addition polymerization

1.2.1 *General features*

A wide variety of olefinic and diolefinic (diene) monomers, e.g. ethylene, styrene, acrylonitrile, vinyl chloride, 1,3-butadiene and isoprene, can be polymerized to give high molecular-weight polymers by radical means. The simplest mechanism of radical polymerization is outlined in Scheme 1.1. The generation of radicals to initiate polymerization is usually accomplished by the thermal or photochemical decomposition of an azo-compound, a peroxide or some other labile additive; alternatively, radical generation may occur spontaneously (as can be the case for example with styrene), be the result of a single electron transfer (redox) reaction, or arise from direct photolysis or radiolysis of the monomer. Once initiation has occurred, propagation is rapid; the average lifetime of a growing radical between initiation and termination under normal conditions is of the order of one second. Termination may be bimolecular by the mutual destruction of growing chains, either by combination or by disproportionation (H-atom transfer), or may involve interaction of the growing radical with another species such as a molecule of solvent or a deliberately added terminating agent. It is obvious even from this much simplified description that radical polymerization involves several random processes and thus offers rather limited possibilities for the control of polymer molecular weight, molecular weight distribution and microstructure. In fact, polymers made by radical polymerization have polydispersities ($\bar{M}_w/\bar{M}_n$ values) of at least 1.5 and are usually atactic (have random stereochemistry), although there may be a slight bias towards syndiotacticity

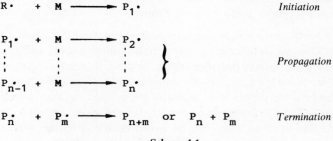

Scheme 1.1

in the case of polymers made from monomers with bulky substituents attached to the alpha carbon atom (e.g. methyl methacrylate).

The control of average molecular weight in radical polymerization is commonly achieved through the use of transfer agents. Such agents usually possess a labile hydrogen atom which can be transferred to the growing radical, thus terminating it. The ideal transfer agent will also reinitiate polymerization with high efficiency and thus its use will not lead to significant reduction in the rate of polymerization. Typical transfer agents are aliphatic thiols (Scheme 1.2). Recent work has highlighted several novel types of additive which serve not only as chain transfer agents but also as initiators and as terminating agents. These agents have been termed 'iniferters'.[4] Their structures and reactions are considered below.

$$P_n^{\cdot} + RSH \longrightarrow P_n\text{-}H + RS^{\cdot} \qquad \text{Transfer}$$

$$RS^{\cdot} + M \longrightarrow P_1^{\cdot} \text{ etc.} \qquad \text{Reinitiation}$$

Scheme 1.2

1.2.2 Iniferters

An iniferter can undergo either spontaneous or induced decomposition to yield a pair of radicals, one of which may initiate polymerization but the other of which is relatively stable and can only take part in chain termination by combining with the growing macroradical, i.e. take part in primary radical termination. The end-group formed by such termination is labile and can dissociate to allow renewed propagation; polymers terminated in such a way therefore have a 'living' character similar to living anionic and cationic chains. Many iniferters also take part in chain transfer reactions, in which case the end-groups introduced during transfer are indistinguishable from those introduced by initiation and primary radical termination. A typical overall mechanism is shown in Scheme 1.3.

Iniferters may be decomposed either photochemically or thermally. Typical of the former type are disulphides such as diphenyl disulphide[5] and dithiuram

$$X\text{-}Y \rightleftharpoons X^{\cdot} + Y^{\cdot} \qquad \text{Dissociation of iniferter}$$

$$X^{\cdot} + M \longrightarrow P_1^{\cdot} \qquad \text{Initiation}$$

$$P_n^{\cdot} + Y^{\cdot} \rightleftharpoons P_nY \qquad \text{Termination}$$

$$P_m^{\cdot} + X\text{-}Y \longrightarrow P_mY + X^{\cdot} \qquad \text{Transfer}$$

Scheme 1.3

disulphide (**1**)[6], and thiocarbamates such as (**2**)[7]. The thiocarbamates offer a distinct advantage over the disulphides in that decomposition of the C–S bond gives a reactive C-centred radical which can initiate polymerization and a less reactive S-centred radical to take part in primary radical termination. The disulphides on the other hand give two identical radicals on decomposition, both of which can initiate polymerization, thus reducing the concentration of S-centred radicals available for termination and increasing the possibility of conventional bimolecular termination between pairs of macroradicals.

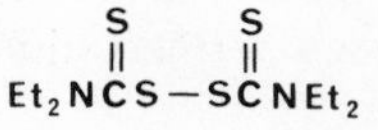

$$Et_2NC(=S)-S-C(=S)NEt_2 \qquad PhCH_2-S-C(=S)NEt_2$$

(1) 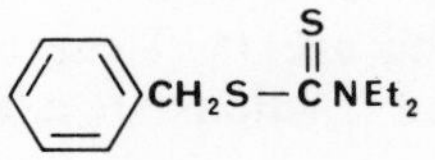 (2)

Typical thermally decomposable iniferters are various pinacol derivatives such as (**3**)[8]. Symmetrical iniferters such as these suffer from the same disadvantage as the disulphides, i.e. both radicals are capable of initiation. A further disadvantage is that some primary radical termination can occur by disproportionation, thus leading to a loss of living ends.[9] (Note: the living end in these cases is the end-group attached to the substituted alpha carbon of the polymer repeat unit; it is the combination of steric compression around this end-group and the stability of the macroradical formed when this end-group decomposes that renders it the more labile of the two.)

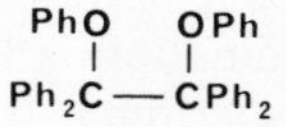

$$Ph_2C(OPh)-C(OPh)Ph_2$$

(3) 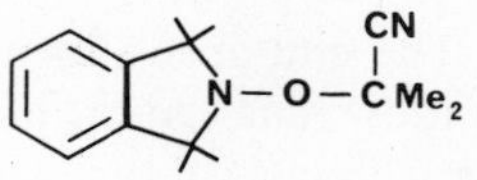 (4)

An interesting and recently described type of iniferter is the alkoxyamine (e.g. (**4**)) formed by decomposing a source of alkyl radicals such as azoisobutyronitrile in the presence of a nitroxide.[10] The C–O bond of the alkoxyamine can undergo reversible thermolysis to yield an alkyl radical capable of initiating polymerization and a stable nitroxyl radical capable only of primary radical termination, unlike the radicals derived from other iniferters.

In theory, the use of iniferters in radical polymerizations allows precise control of average chain lengths. However, in practice, the molecular weights achieved with such agents are rather low. The major, but as yet largely potential, use of iniferters is for the construction of block copolymers, since a homopolymer constructed in the presence of an iniferter is a macroinitiator and can be used to initiate the polymerization of a second monomer.[6,11,12] This aspect is considered further in Chapter 6.

1.2.3 Template polymerization

Template polymerization (the construction of a daughter polymer on a template of a preformed polymer) is a well-recognized phenomenon in natural systems, e.g. in the self-replication of DNA and in the biosynthesis of proteins. It is not surprising therefore that the possibilities of a template effect should have been investigated also for synthetic polymers and especially for those made by radical polymerization where control of polymer molecular weight and microstructure is most difficult to achieve.[13,14]

Template effects can be expected to occur in a situation where two different polymer chains can form an interpolymer complex, and the identifying of such complexes between preformed polymers can be a useful way of screening for the possibility of such an effect. Template effects in polymerization may be revealed in one or more of the following ways:

1. by an enhanced rate of polymerization in the presence of the template;
2. by correlations between the molecular weight of the daughter polymer and that of the template; and
3. by differences between the microstructure of the daughter polymer and that of a similar polymer made in the absence of the template.

Radical template polymerizations have been classified into two main types: Type 1, in which the monomer for the daughter polymer is strongly pre-adsorbed on the template polymer, and Type 2, in which polymerization of the daughter monomer begins in the bulk of the solution and adsorption of the growing chain on the template polymer occurs only when a critical size has been reached (see Scheme 1.4). Type 1 systems in general are those where there are strong ionic or charge-transfer interactions between the template and daughter monomer/polymer, whereas in Type 2 systems the interactions are of a weaker hydrogen-bond type.

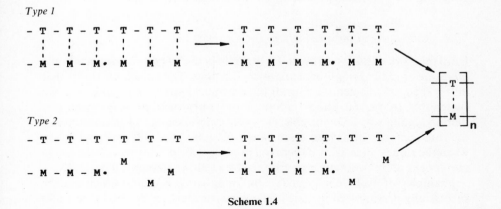

Scheme 1.4

Examples of Type 1 systems include polymerizations of acrylic acid, methacrylic acid and *p*-styrenesulphonic acid on templates such as polyethyleneimine and poly-L-lysine and polymerizations of vinyl pyridines on various polyacid templates. It is in polymerizations such as these where the greatest effects of template upon the structure of the daughter might be expected. However, complete separation of the daughter polymer from the template is often difficult when the interactions between them are strong and only a few systems have been characterized in detail from the structural point of view. It is interesting to note, however, that there are some reports of correlations between molecular weights of daughter and template polymers[15-17] and that in one system at least (the polymerization of vinyl sulphonate on an ionene template) the daughter polymer was more isotactic than a similar polymer made in the absence of the template.[18]

Type 2 behaviour is exemplified by polymerizations of *N*-vinylpyrrolidone in the presence of tactic poly(acrylic acid)[19] or poly(methacrylic acid)[20,21] in DMF, by the converse type of reaction[22] and also by methyl methacrylate along isotactic or syndiotactic poly(methyl methacrylate).[13] The latter system is an example of one in which the forces of interaction, although only of a weak Van der Waals type, are strong enough to cause the daughter polymer to have a tacticity complementary to that of the template (syndiotactic daughter polymer formed on isotactic template and vice versa).

Clearly, then, template polymerizations can offer some control of polymer molecular weight and stereochemistry in selected cases. However, for a variety of reasons, not least the difficulty often experienced in separating daughter polymer from the template, practical applications have been slow to emerge. Thus the few applications reported so far tend to use the interpolymer complex arising from template polymerization directly in the final product, for example, the use of acrylic acid/poly(ethylene oxide) template systems as precursors in the formation of interpenetrating polymer networks.[23,24] Problems of separating daughter polymer from matrix could be reduced by attaching the template polymer to an inert, insoluble matrix.[25]

1.2.4 *Ring-opening polymerization*

Until the work of Bailey through the 1980s[26] there had been few reports in the literature of radical ring-opening polymerizations. The reason for this is that unless there is considerable ring strain, for many cyclic compounds that are susceptible to radical attack (addition or H-abstraction), ring-opening is thermodynamically unfavourable. However, ring-opening can be favourable if the ring contains an oxygen atom which becomes part of a carbonyl group when the ring is opened, or if there is a more specific driving force such as the attainment of an aromatic structure in the resulting polymer repeat unit.

Examples of cyclic compounds containing oxygen atoms which can be successfully polymerized include cyclic ketene acetals, cyclic vinyl ethers and cyclic α-alkoxyacrylates. 2-Methylene-1,3-dioxolane (5) undergoes partial ring-opening on polymerization at temperatures of up to 120 °C to give a polymer containing both in-chain ester links and pendant rings (equation 1.1)[27], whereas 2-methylene-1,3-dioxepane (6) undergoes complete ring-opening polymerization even at room temperature.[28] Nitrogen and sulphur analogues of cyclic ketene acetals will also undergo ring-opening polymerization.[26] In the case of the nitrogen analogue, ring-opening is complete even at low temperature for the five-membered ring compound because of the low enthalpy of the resulting amide group; for the five-membered ring sulphur analogue, however, only 45% ring-opening was achieved at 160 °C, indicating the somewhat higher enthalpy of the thioester group.

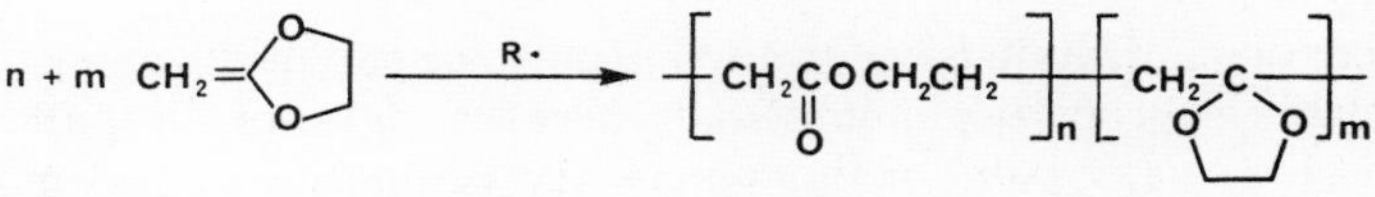

$$n + m \ CH_2{=}\!\!\bigcirc \xrightarrow{R\cdot} {-}[CH_2\underset{\underset{O}{\|}}{C}OCH_2CH_2]_n[CH_2{-}\underset{O\diagdown\diagup O}{C}]_m{-} \quad (1.1)$$

(5)

For cyclic vinyl ethers, ring-opening is also incomplete, even in cases where considerable ring-strain must be present and where groups are added to help stabilize the intermediate radicals produced (e.g. phenyl groups) (7).[29] Cyclic α-alkoxyacrylates are also reluctant to undergo ring opening although monomer (8) gave a polymer containing about 50% open-chain units at 140 °C;[30] they are, however, rather more reactive in radical polymerization than cyclic ketene acetals and cyclic vinyl ethers.

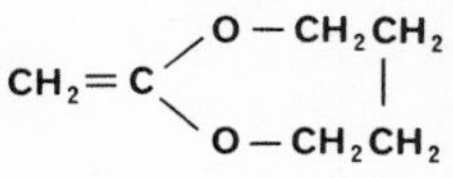

$$CH_2{=}C\diagup^{O-CH_2CH_2}_{O-CH_2CH_2} \qquad Ph\diagdown\!\!\bigcirc\!\!{=}CH_2 \qquad Ph\diagdown\!\!\bigcirc\!\!{=}CH_2$$

(6) 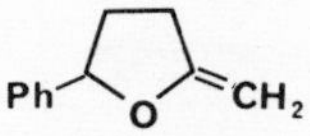   (7) 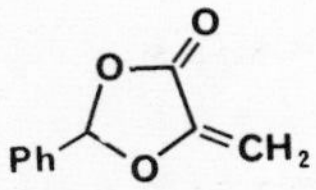   (8)

Methylene spirohexadienes such as (9) are examples of monomers for which the driving force for ring-opening polymerization is largely the formation of an aromatic ring in the monomer repeat unit (equation 1.2).[31]

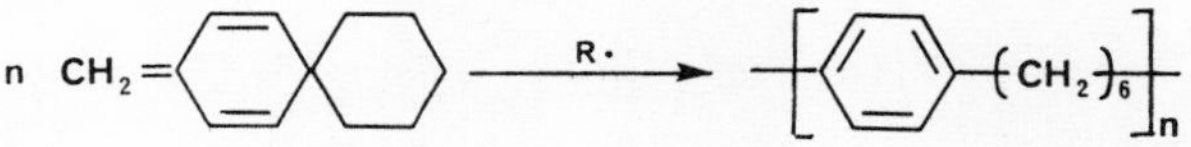

$$n \ CH_2{=}\!\!\bigcirc\!\!\!\bigcirc \xrightarrow{R\cdot} {-}[\!\bigcirc\!{-}(CH_2)_6]_n{-} \quad (1.2)$$

(9)

Ring-opening radical polymerizations as well as being of intrinsic interest have been shown also to have practical applications. Cyclic ketene acetals, for example, can be copolymerized with a variety of comonomers. When a small amount of 2-methylene-1,3-dioxepane (6) is copolymerized with ethylene a polymer is produced which is significantly more biodegradable than polyethylene owing to the presence in the polymer backbone of ester groups.[32] Copolymers of (6) with methyl methacrylate are significantly more thermally stable than poly(methyl methacrylate) itself, presumably because the depolymerization of the methyl methacrylate sequences is blocked by the presence of the additional ester links.[26] Copolymers containing ester links derived from cyclic ketene acetals can also be hydrolysed to give terminally functionalized oligomers (see Chapter 6). Some cyclic monomers, e.g. unsaturated spiro orthocarbonates, polymerize with no diminution in volume or even a slight increase in volume; such monomers have potential uses in dental filling materials.[33]

1.2.5 Copolymerization

Radical copolymerization is widely used to produce polymers having physical and mechanical properties intermediate between those of the pure homopolymers. For many pairs of monomers, the compositions and monomer sequence distributions of the copolymers produced can be rationalized if it is assumed that the rate constant for a particular propagation step depends only upon the nature of terminal monomer unit in the attacking radical and on the chemical nature of the monomer which is being attacked, i.e. in terms of just four propagation steps (Scheme 1.5).[34] For such a mechanism, the

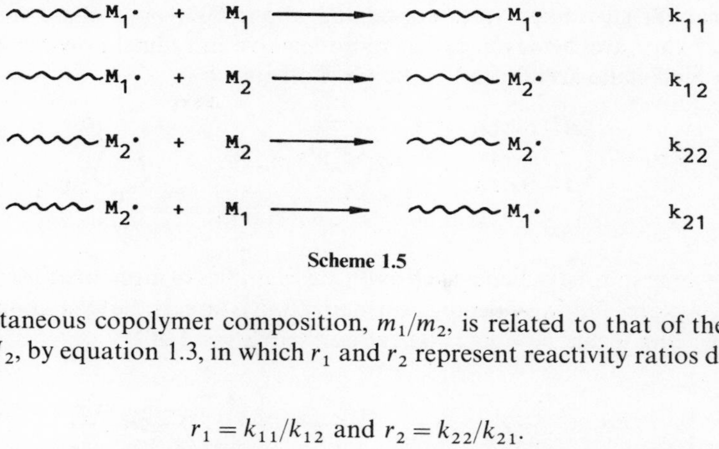

$$\sim\sim\sim M_1\cdot\ +\ M_1\ \longrightarrow\ \sim\sim\sim M_1\cdot\qquad k_{11}$$

$$\sim\sim\sim M_1\cdot\ +\ M_2\ \longrightarrow\ \sim\sim\sim M_2\cdot\qquad k_{12}$$

$$\sim\sim\sim M_2\cdot\ +\ M_2\ \longrightarrow\ \sim\sim\sim M_2\cdot\qquad k_{22}$$

$$\sim\sim\sim M_2\cdot\ +\ M_1\ \longrightarrow\ \sim\sim\sim M_1\cdot\qquad k_{21}$$

Scheme 1.5

instantaneous copolymer composition, m_1/m_2, is related to that of the feed, M_1/M_2, by equation 1.3, in which r_1 and r_2 represent reactivity ratios defined by

$$r_1 = k_{11}/k_{12} \text{ and } r_2 = k_{22}/k_{21}.$$

Equations involving r_1 and r_2 can also be written which describe the way in which the copolymer composition will vary with time and which give the fractions of various monomer sequences within the copolymer as a function of feed or copolymer composition. Since so much depends upon r_1 and r_2, considerable effort has been expended in trying to understand the aspects of

$$\frac{m_1}{m_2} = \frac{M_1}{M_2}\left[\frac{r_1 M_1 + M_2}{r_2 M_2 + M_1}\right] \qquad (1.3)$$

monomer and radical structure and reactivity which determine them. For a given pair of monomers, considerable advantage could accrue if reactivity ratios could be altered in a predictable way to generate copolymers with different sequence distributions but with similar compositions.

Unfortunately, reactivity ratios are rather insensitive to temperature and also, with some exceptions,[35,36] rather insensitive to the nature of the medium. Nevertheless, it has been demonstrated that, for certain pairs of monomers, reactivity ratios can be much reduced and hence highly alternating copolymers be produced by addition of small quantities of metal halide, or organometal halide, complexing agents.[37] Typical systems are those in which one of the monomers is of a donor type, e.g. styrene, vinyl acetate, butadiene and isoprene, and the other is an acceptor, e.g. methyl acrylate, methyl methacrylate, acrylonitrile or methyl vinyl ketone. A further requirement is that the acceptor monomer should possess a carbonyl or nitrile group capable of complexing with the metal or organometal halide and which is conjugated with the olefinic double bond. Complexing agents which have been employed include $ZnCl_2$, BCl_3, $SnCl_4$, Et_2AlCl, $EtAlCl_2$ and $Et_3Al_2Cl_3$. Initiation of copolymerization may be spontaneous (especially at higher temperatures) or may be induced by the thermal or photochemical decomposition of an added peroxide or azo-compound. Although copolymerization in the presence of complexing agents is a radical process, the exact way in which alternation is enhanced is still the subject of some dispute. Proper analysis of some systems is complicated by a concurrent cationic polymerization of the donor monomer initiated by the complexing agent.

The practical advantages of being able to increase alternation in some copolymerizations appear to be relatively few, although the production of a synthetic elastomer based on an alternating acrylonitrile–butadiene copolymer has been described.[38] Nevertheless, other applications can be envisaged, especially in situations where the highly regular structure of an alternating copolymer could be exploited, such as in making highly regular Langmuir–Blodgett films with potential opto- and micro-electronic uses and as components of high definition resists.

Some control of copolymer microstructure may also be available using template methods similar to those mentioned above. A few such systems have been investigated[13,15,16,39,40] including some in which block copolymers of styrene with methacrylic acid and of acrylonitrile with methacrylic acid appear to have been produced, albeit by indirect routes.[41,42]

1.3 Ionic and coordination addition polymerization

1.3.1 *General background and developments*

Although used less extensively in industry than radical addition and stepwise polymerization, ionic and coordination polymerization processes are nevertheless important for the large-scale production of several polymers.

Anionic polymerization in particular is used to produce cis-1,4-polyisoprene, styrene-diene block copolymers, and to make castings of nylon-6 (poly-ε-caprolactam). The formation of the block copolymers typically uses butyl lithium to initiate a living polymerization of styrene; sequential additions of the diene and styrene then produce alternating blocks. The anionic polymerization of ε-caprolactam has attracted interest recently for applications in reaction injection moulding (RIM); in this process, molten ε-caprolactam is injected into a mould with a suitable initiator (e.g. an acyllactam), a catalyst (an alkali, or alkaline earth, metal or derivative) and usually also an α,ω-lactam-terminated oligoether or oligoester. The resulting material is a block copolymer containing alternating oligoamide and oligoether (or oligoester) blocks (Scheme 1.6).[43]

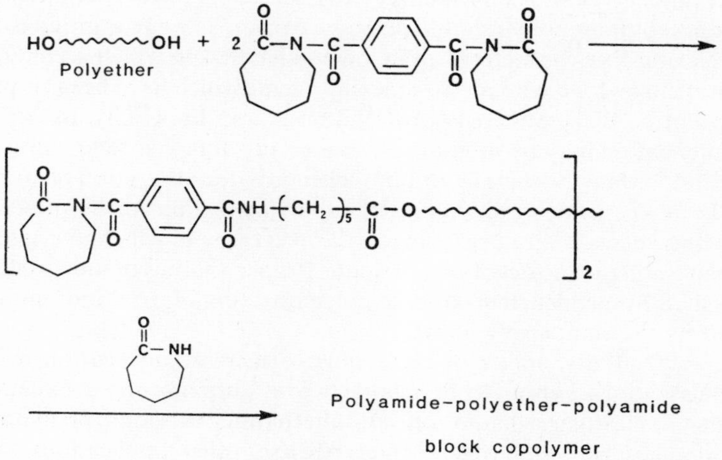

Polyamide–polyether–polyamide block copolymer

Scheme 1.6

The current major uses of cationic polymerization are for the production of polyisobutene, butene/diene copolymers, acetal resins (by the cationic homo- and copolymerization of 1,3,5-trioxane (**10**) and polytetrahydrofuran. Polytetrahydrofuran is largely produced as an α,ω-dihydroxy telechelic oligomer, in which form it is used extensively as a soft segment in the construction of segmented polyurethanes.

Coordination catalysis, and especially Ziegler–Natta and related processes, offers unrivalled control of stereochemistry in the polymerization of alkenes and dienes. In particular, coordination methods are used on a large scale to produce high density polyethylene, isotactic polypropylene, isotactic poly(but-1-ene), ethylene-propylene co- and ter-polymers, linear low density polyethylene (made by copolymerizing ethylene with a small amount of but-1-ene), cis-1,4-polybutadiene, and cis-1,4- and trans-1,4-polyisoprenes.

Much of the recent research effort in ionic and coordination polymerization has been directed towards extending the utility of the techniques, particularly towards the development of methods capable of accomplishing the ring-opening polymerization of a variety of cyclic compounds, thus giving access to a wider range of polymer structures, and the development of true 'living' systems which offer optimum control of molecular weight and molecular weight distribution.

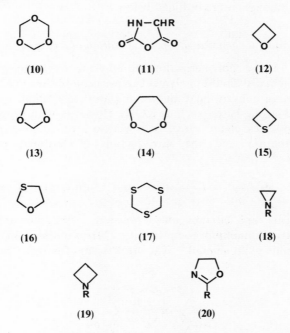

Ring-opening polymerizations of cyclic alkenes and dienes (metathesis polymerization), which can be accomplished using a variety of W–, Mo–, Ru– and Re–based coordination catalysts, is covered in Chapter 3. Epoxides (oxiranes), episulphides (thiiranes), lactones, lactams, N-carboxyanhydrides (11) and cyclosiloxanes are examples of cyclic compounds, many of which have been shown to be polymerizable by an anionic mechanism. Several of the above can also be polymerized cationically, in fact the utility of the cationic technique is rather greater since a wider range of heterocyclic monomers can be polymerized by this mechanism including oxetanes (12), oxolanes (e.g. THF) and other cyclic ethers, cyclic acetals such as 1,3-dioxolane (13) and 1,3-dioxepane (14) (as well as 1,3,5-trioxane mentioned above), and a variety of S- and N-containing heterocycles such as thietane (15), 1,3-oxathiolane (16), 1,3,5-trithiane (17), aziridines (18), azetidines (19) and 1,3-oxazolines (20). Some of these anionic and cationic ring-opening polymerizations are considered in Chapter 6 but developments in these areas have been comprehensively reviewed.[44–46]

One of the most significant recent developments in living polymerization has been the discovery of what has been termed 'group transfer polymerization' or GTP.[47] GTP can be used on various acrylic monomers and is initiated by silyl ketene acetals in the presence of either Lewis acid catalysts or weakly nucleophilic anions. GTP resembles anionic polymerization in many respects; it is discussed in detail in Chapter 2. Three other recent developments in living polymerization are outlined below.

1.3.2 Living anionic homo- and copolymerization of oxiranes

The 'living' anionic polymerization of ethylene oxide to give narrow molecular-weight distribution polymer can be accomplished in aprotic media with a variety of conventional anionic initiators, e.g. alkali metals, metal alkoxides and strong bases (such as KOH). However, with propylene oxide and higher epoxides, chain transfer can lead to a broadening of molecular weight distribution and the introduction of terminal unsaturation (equation 1.4).[48]

$$\sim\!\!\sim O^- \; M^+ \; + \; Me\!-\!\!\triangle_O \longrightarrow \sim\!\!\sim OH \; + \; CH_2\!=\!CHCH_2O^- \; M^+ \quad (1.4)$$

These problems are circumvented, however, with the use of tetraphenylporphyrin–aluminium complexes (e.g. (**21**)) as initiators in the presence of small amounts of an alcohol.[49] The mechanism of polymerization can be

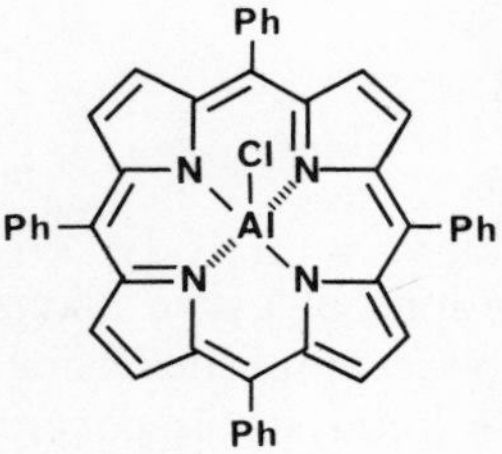

(**21**)

represented as a repeated ring-opening insertion of ethylene oxide into the Al–O bond, formed initially by reaction of the complex with the alcohol (Scheme 1.7). The resulting polymers are of very narrow molecular-weight

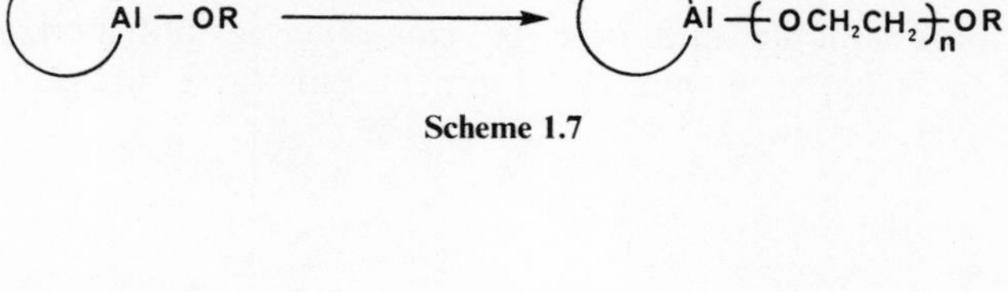

Scheme 1.7

DEVELOPMENTS IN POLYMERIZATION 13

distribution (typically $\bar{M}_w/\bar{M}_n = 1.05$) and have been dubbed 'immortal' since they retain living character even in the presence of proton donors such as HCl, carboxylic acids and alcohols.[50,51] Block copolymers can be made from the 'immortal' polyethers with little broadening of molecular weight distribution, for example, by introducing ε-caprolactone which is ring-opened at the reactive Al end-group to form a polyester block.[51]

1.3.3 *The activated monomer mechanism of oxirane polymerization*

It is well known that in the conventional cationic ring-opening polymerization of oxiranes (e.g. initiation by BF_3 or PF_5 in aprotic media), propagation is accompanied by some cyclization of the growing chain-end and the consequent formation of various cyclic oligomers (Scheme 1.8). Indeed, this

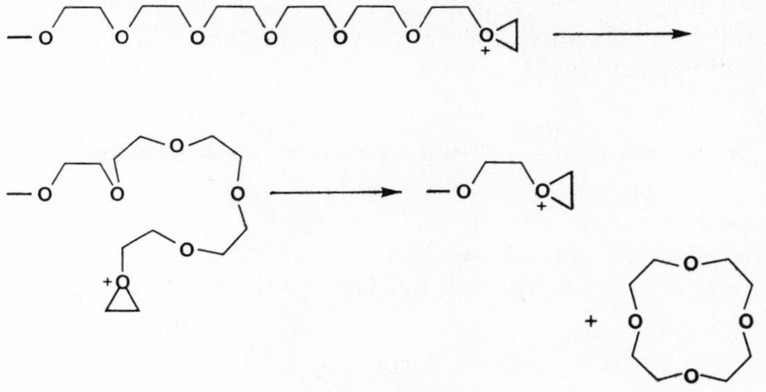

Scheme 1.8

cyclization has been utilized to make crown ethers.[52] Recently, however, it has been demonstrated that the formation of cyclic species can be much reduced by carrying out the polymerization in the presence of small amounts of an added alcohol.[53] This mechanism, known as the activated monomer mechanism, is believed to involve the attack of a neutral nucleophilic chain-end upon a protonated, and therefore activated, monomer with the consequent release of a proton which protonates a further molecule of monomer (Scheme 1.9).[54] It can be said, therefore, that the alcohol is the true initiator and that the proton acts as a catalyst.

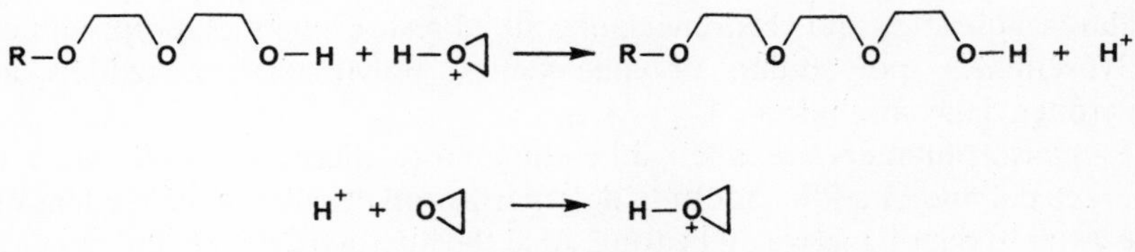

Scheme 1.9

1.3.4 *Living carbocationic polymerization of vinyl ethers*

The cationic polymerization of vinyl ethers can be initiated by a variety of acidic species including protonic acids, and Lewis acids (e.g. $SnCl_4$, BF_3, $FeCl_3$, $RAlCl_2$ and $RMgX$, where R = alkyl and X = halogen). The polymers so formed usually have broad molecular weight distributions owing to transfer of a β-proton from the active chain end to the incoming monomer.

However, it has been found recently that vinyl ethers can be polymerized at low temperature by a mixture of hydrogen iodide and iodine to give living polymers with narrow molecular weight distributions ($\bar{M}_w/\bar{M}_n < 1.1$) regardless of conversion.[55,56] The living polymers can be further converted to block copolymers or be end-functionalized to form telechelic polymers. The mechanism of polymerization is thought to involve the preliminary formation of a vinyl ether/HI adduct which acts as an initiator in the presence of I_2 (Scheme 1.10).[57] The process promises to be particularly useful for the preparation of poly(vinyl ethers) carrying pendant functional groups for specialized applications.[58]

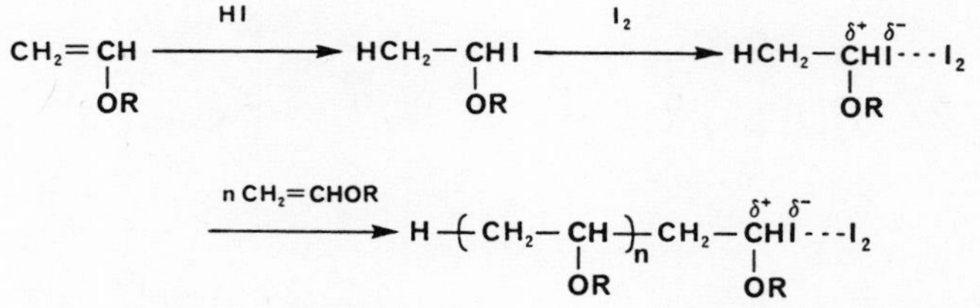

$$CH_2=CH\underset{OR}{|} \xrightarrow{HI} HCH_2-CHI\underset{OR}{|} \xrightarrow{I_2} HCH_2-\overset{\delta^+}{C}\overset{\delta^-}{HI}\cdots I_2\underset{OR}{|}$$

$$\xrightarrow{n\,CH_2=CHOR} H\text{-}(CH_2-CH\underset{OR}{|})_n CH_2-\overset{\delta^+}{C}\overset{\delta^-}{HI}\cdots I_2\underset{OR}{|}$$

Scheme 1.10

1.4 Stepwise polymerization

1.4.1 *General background*

Stepwise polymers (polymers made by condensation and rearrangement processes) as a class include most of the polymers currently used as strong and/or tough plastics and fibres, such as polyesters, polycarbonates and polyamides, and all of the important thermosetting resins such as epoxy resins, unsaturated polyester resins, amino resins and phenolic resins. They also include many materials used in particular to make speciality coatings, sealants, adhesives and elastomers, like alkyd resins, silicones, polysulphides, polyurethanes, polyimides, polyphenylenes, poly(phenylene oxide)s and poly(phenylene sulphide)s.

Stepwise polymers are often inherently more thermally stable than the average polymer made by addition polymerization because of the tendency for the latter to depolymerize on heating. Also the microstructures, and hence the chemical, physical and, to a lesser extent, mechanical properties, of stepwise polymers are easier to predict from the structures and properties of the constituent monomers. It is not surprising therefore that considerable attention should have been focussed more recently on stepwise processes as a means of producing high-performance polymers, for example polymers with high thermal stability and/or mechanical properties suitable for their application as engineering materials. Some developments in these areas are described below.

1.4.2 Aromatic polyesters and polyamides

Commercial production of aromatic polyesters, or polyarylates as they are often called, began in 1970 with the introduction (by the Carborundum Company) of the highly crystalline poly(4-hydroxybenzoic acid) (**22**), followed a few years later by amorphous polymers made by condensing bisphenol A with a mixture of tere- and iso-phthalic acids (**23**). Polymers with varying degrees of crystallinity intermediate between those of the above two examples can be made by condensing different diphenols with appropriate diacids or their derivatives.

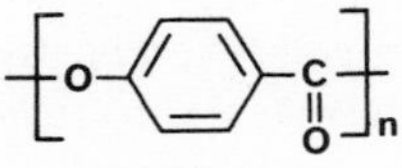

(**22**) 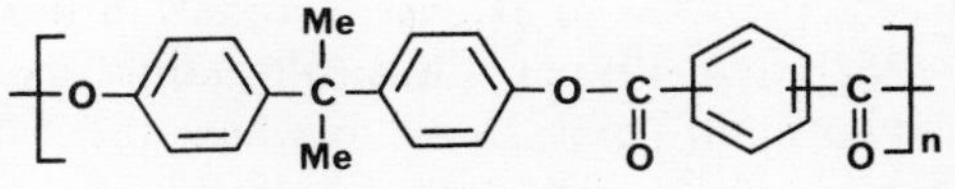   (**23**)

The current methods for making polyarylates involve not the direct condensation of aromatic diol with aromatic diacid but reaction between the diol and the diacid chloride, either in an interfacial or in a solution process, or reaction between a diacetate and a diacid, or reaction between a diphenate and a diol. These alternative strategies involve the elimination of HCl, acetic acid and phenol respectively in the condensation process. Polyarylates are noted for their high thermal and UV stability. It should be noted, however, that many polyarylates can undergo a photo-Fries rearrangement (equation 1.5) with consequent yellowing, although the rearranged polymer is subsequently UV stable.[59,60]

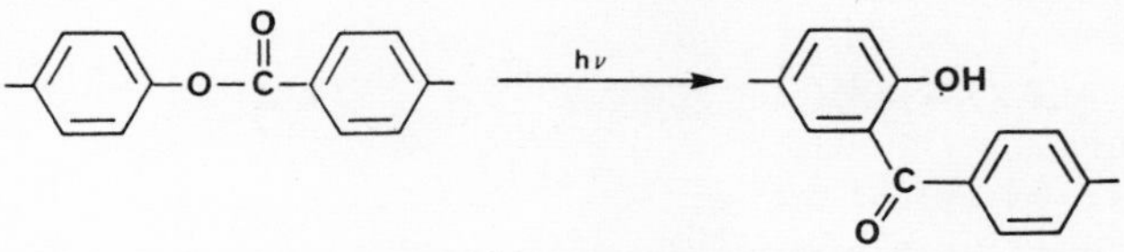

(1.5)

The first commercially produced wholly aromatic polyamide was poly(*m*-phenyleneisophthalimide) (**24**), introduced under the tradename of Nomex by DuPont in 1961. This was shortly followed by poly(*p*-phenyleneterephthalimide) (**25**) marketed under the tradenames of Kevlar and Twaron.

$\left[NH-\bigcirc-NHC-\bigcirc-\underset{O}{\overset{C}{\underset{\|}{C}}}\right]_n$ $\left[NH-\bigcirc-NHC-\bigcirc-\underset{O}{\overset{C}{\underset{\|}{C}}}\right]_n$

(24)   (25)

The polymers are made by reaction of the appropriate diamine with the appropriate diacid chloride in a low temperature interfacial or solution polycondensation process. Both polymers are mainly used for making fibres; fibres from the former have important uses in making flame and chemically resistant materials, while those from the latter are used extensively for the reinforcement of rubbers and plastics. The *para*-coupled polyamides adopt rod-like conformations in solution which, when the solution is sufficiently concentrated, align to give (lytropic) liquid–crystalline hebaviour. These alignments persist in fibres spun from such solutions and contribute to their tensile strength. Similar (thermotropic) liquid crystalline behaviour can be displayed by melts of such polymers, and of the similarly coupled polyarylates; these regions of alignment persist on cooling, the polymer giving rise to a self-reinforcing effect and thus improving mechanical properties.

The observation of liquid crystalline behaviour in aromatic polyamides and polyesters has led to extensive studies of this phenomenon covering a wide range of polymer types.[61] Basically, liquid crystalline polymers can be divided into two main types: those in which the inflexible (mesogenic) groups are incorporated in the polymer backbone and those in which they are incorporated as side-chains (Figure 1.1). Liquid crystalline polymers have potential applications not only for making strong fibres and tough plastics but also in more specialised areas such as in liquid crystal displays and as information storage media.

1.4.3 *Polysulphones and polyketones*

Aromatic polymers containing sulphone and ketone linkages in the main-chain are attracting great and increasing interest as engineering thermo-

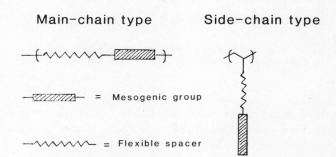

Figure 1.1 Schematic representation of typical liquid crystalline polymers

plastics. In order to produce tractable materials, it is usual also to incorporate ether linkages as well; (**26**) and (**27**) are two typical commercially available materials (from Union Carbide and ICI respectively). Aromatic polysulphones can be made by a number of routes including by polysulphonylation using Friedel–Crafts catalysts. However, the synthesis of poly(ether sulphone)s such as (**26**) can be accomplished more readily by polyetherification involving the

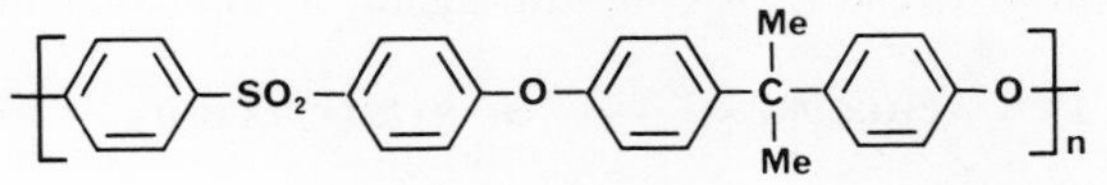

(**26**)

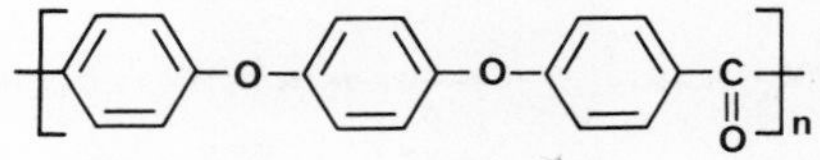

(**27**)

nucleophilic displacement of halogen from a preformed sulphone by phenoxide (equation 1.6).[62].

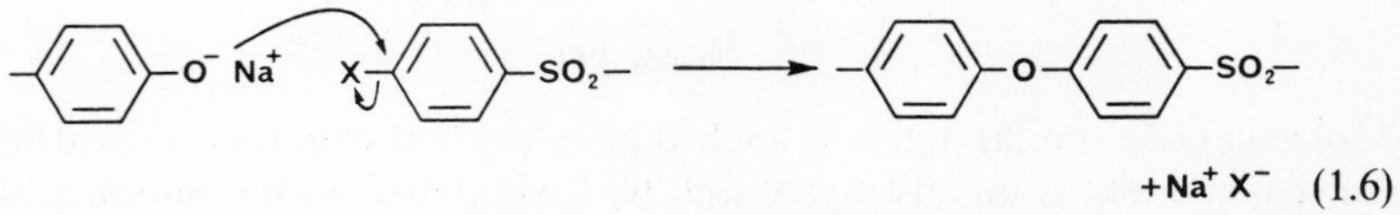

$+Na^+ X^-$ (1.6)

Poly(ether ketone)s are made by similar chemistry although in these cases the displaced group is F^- or NO_2^- rather than Cl^-;[63] (**27**), often referred to as PEEK, is usually made from hydroquinone and 4,4'-difluorobenzophenone. The chemistry clearly lends itself to the synthesis of a wide variety of materials including hybrids possessing ether, sulphone and ketone linkages.

Whilst the commercial polysulphones are amorphous polymers, the polyketones are not. PEEK is partly crystalline and has a melting point (T_m) of 335 °C; its T_g is 144 °C.[64] The high T_m means that processing of PEEK must be carried out at high temperatures, but nevertheless it contributes significantly to the good high temperature performance of the polymer. PEEK is being investigated particularly as a matrix material for fibre-reinforced composites and also as a high-performance fibre.

1.4.4 Starburst polymers: dendrimers and arborols

The step-by-step construction of a synthetic high molecular weight polymer that is perfectly monodisperse and that has a precisely predetermined microstructure (as is the case with biopolymers such as polypeptides and nucleic acids) has been a long-term goal. This goal has recently been achieved in the multistage synthesis of a number of highly branched polymers using repeated condensation and/or rearrangement processes. The first such example is the synthesis of what have been termed 'dendrimers', or 'starburst' polymers, by reactions involving repeated Michael additions between amines and acrylates or methacrylates;[65–67] one such methodology, based on Michael additions of ammonia (or another amine) to methyl acrylate and using 1,2-diaminoethane as a coupling agent, is indicated in Scheme 1.11.

$$NH_3 \; + \; 3\,CH_2{=}CHCO_2Me \longrightarrow N(CH_2CH_2CO_2H)_3$$

(28) (29)

$$(29) \; + \; 3\,NH_2CH_2CH_2NH_2 \longrightarrow N(CH_2CH_2CONHCH_2CH_2NH_2)_3$$

(30)

$$(30) \; + \; 6\,(28) \longrightarrow N(CH_2CH_2CONHCH_2CH_2N(CH_2CH_2CO_2Me)_2)_3$$

Dendrimer

Scheme 1.11

Molecules of a similar highly branched nature called 'arborols' have also been constructed, but using aldehydes and hydroxyamines as the building blocks (e.g. (**31**)).[68,69] Dendrimers can behave as spherical micelles[70,71] and have potential applications as biocatalytic mimics[66] and as drug-delivery agents.[72]

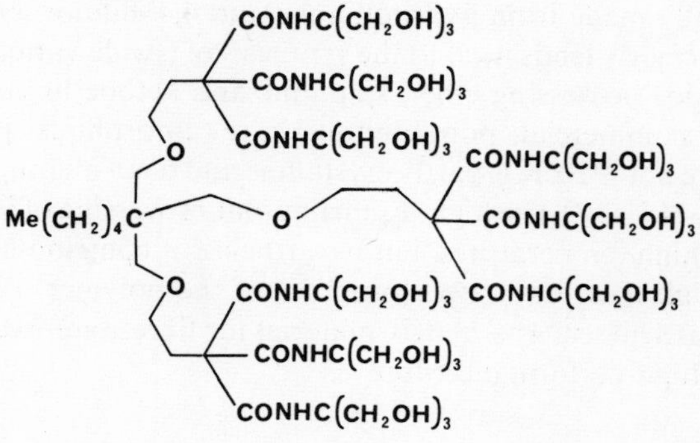
CONHC(CH2OH)3
CONHC(CH2OH)3
CONHC(CH2OH)3
O
CONHC(CH2OH)3
Me(CH2)4C
O
CONHC(CH2OH)3
O
CONHC(CH2OH)3
CONHC(CH2OH)3
CONHC(CH2OH)3
CONHC(CH2OH)3

(31)

1.4.5 *Biosynthetic routes to polymers*

Nature synthesizes a variety of polymers by what are essentially stepwise methods, including polysaccharides, proteins and polyisoprenes. Several of these polymers are already heavily exploited by man and have been for many decades, e.g. regenerated cellulose and cellulose acetate as fibres, cellulose nitrate as a surface coating, and *cis*-1,4-polyisoprene as an elastomer. Obtaining other useful polymers from living organisms, either by genetic engineering or by providing a suitable feedstock, is a well-recognized possibility but as yet there have been very few developments. Of particular note, however, is the exploitation by ICI of a bacterial route to poly(β-hydroxybutyrate) (32).

$$\left[\text{OCHMeCH}_2\underset{\underset{O}{\|}}{C}\right]_n$$

(32)

By suitable adjustments to feedstocks it is possible also to persuade the bacteria to produce other poly(β-hydroxyalkanoate)s[73] and also to produce random copolymers.[74,75] The polymers are of very high molecular weight ($>10^6$), have good thermal stability and are biodegradable; attributes which could make them increasingly important in the future.

Also worthy of note here, although it is a precursor to an addition polymerization, is the preparation of the *cis*-diacetate derivative of 5,6-dihydroxy-1,3-cyclohexadiene by the genetically engineered bacterium *Pseudomonas putida*.[76] Radical polymerization of the diacetate followed by heating at 200–300 °C results in the formation of polyphenylene (Scheme 1.12).

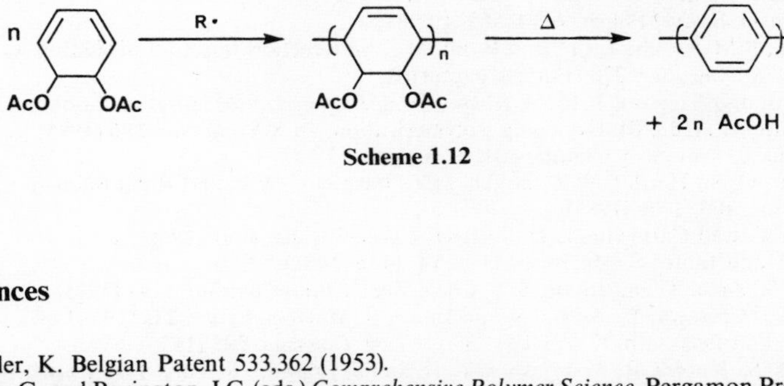

Scheme 1.12

References

1. Ziegler, K. Belgian Patent 533,362 (1953).
2. Allen, G. and Bevington, J.C. (eds.) *Comprehensive Polymer Science*, Pergamon Press, Oxford, (1989).
3. Kroschwitz, J.I. (ed.) *Encyclopedia of Polymer Science and Engineering*, Wiley, New York (2nd edn.), (1985–1989).
4. Otsu, T. and Yoshida, M. *Markromol. Chem., Rapid Commun.* **3**, 127 (1982).
5. Otsu, T., Yoshida, M. and Kuriyama, A. *Polym. Bull.* **7**, 45 (1982).
6. Otsu, T. and Kuriyama, A. *Polym. Bull.* **11**, 135 (1984).
7. Otsu, T. and Kuriyama, A. *J. Macromol. Sci.* **A21**, 961 (1984).
8. Bledzki, A. and Braun, D. *Makromol. Chem.* **182**, 1047 (1981).
9. Otsu, T. and Tazaki, T. *Polym. Bull.* **17**, 187 (1987).
10. Bednarek, D., Moad, G., Rizzardo, E. and Solomon, D.H. *Macromolecules* **21**, 1522 (1988).

11. Kuriyama, A. and Otsu, T. *Polym. J.* **16**, 511 (1984).
12. Otsu, T. and Kuriyama, A. *Polym. J.* **17**, 97 (1985).
13. Tan, Y.Y. and Challa, G. *Makromol. Chem., Macromol. Symp.* **10/11**, 215 (1987).
14. Tan, Y.Y. in Reference 2, Volume 3, Chapter 19.
15. Shima, K., Kakui, Y., Kinoshita, M. and Imoto, M. *Makromol. Chem.* **154**, 247 (1972).
16. Shima, K., Kakui, Y., Kinoshita, M. and Imoto, M. *Makromol. Chem.* **155**, 299 (1972).
17. Miura, Y., Kakui, Y. and Kinoshita, M. *Makromol. Chem.* **176**, 1567 (1975).
18. Blumstein, A., Kakivaya, S.R., Blumstein, R. and Suzuki, T. *Macromolecules* **8**, 435 (1975).
19. Rajan, V.S. and Ferguson, J. *Eur. Polym. J.* **18**, 633 (1982).
20. Bartels, T., Tan, Y.Y. and Challa, G. *J. Polym. Sci., Polym. Chem. Ed.* **15**, 341 (1977).
21. Koetsier, D.W., Tan, Y.Y. and Challa, G. *Polymer* **22**, 1709 (1981).
22. Alberda van Ekenstein, G.O.R. and Tan, Y.Y. *Polym. Commun.* **25**, 105 (1985).
23. Nishi, S. and Kotaka, T. *Macromolecules* **18**, 1519 (1985).
24. Nishi, S. and Kotaka, T. *Macromolecules* **19**, 978 (1986).
25. Sherrington, D.C. and Hodge, P. (eds.) *Synthesis and Separations using Functional Polymers* Wiley, Chichester (1988).
26. Bailey, W.J. in Reference 2, Volume 3, Chapter 22.
27. Bailey, W.J. *Polymer* **17**, 85 (1985).
28. Bailey, W.J., Ni, Z. and Wu, S.-R. *J. Polym. Sci., Polym. Chem. Ed.* **20**, 2420 (1982).
29. Tsang, R., Dickson, Jr., J.K., Pak, H., Walton, R. and Fraser-Reid, B. *J. Amer. Chem. Soc.* **109**, 3484 (1987).
30. Bailey, W.J. and Feng, P.-Z. *Polym. Prepr.* **28**(1), 154 (1987).
31. Bailey, W.J., Amone, M.J. and Chou, J.L. *Polym. Prepr.* **29**(1), 178 (1988).
32. Bailey, W.J. and Gapud, B. *ACS Symp. Ser.* **280**, 423 (1985).
33. Thompson, V.P., Williams, E.F. and Bailey, W.J. *J. Dent. Res.* **58**, 522 (1979).
34. Mayo, F.R. and Lewis, F.M. *J. Amer. Chem. Soc.* **66**, 1594 (1944).
35. Harwood, H.J. *Makromol. Chem., Macromol. Symp.* **10/11**, 331 (1987).
36. Chapiro, A. *Eur. Polym. J.* **7/8**, 713 (1989).
37. Bamford, C.H. in *Alternating Copolymers.* (ed. Cowie, J.M.G.), Plenum Press, New York (1985), Chapter 3 and references cited therein.
38. Takamatsu, T., Ohnishi, O., Nishikida, T. and Furukawa, J. *Rubber Age* **105**, 33 (1973).
39. Polowinski, S. *Eur. Polym. J.* **19**, 679 (1983).
40. Polowinski, S. *J. Polym. Sci., Polym. Chem. Ed.* **22**, 2887 (1984).
41. Polowinski, S. and Janowska, G. *Eur. Polym. J.* **11**, 183 (1975).
42. Polowinski, S. *Eur. Polym. J.* **14**, 563 (1978).
43. Hedrick, R.M., Gabbert, J.D. and Wohl, M.D. in Reaction Injection Moulding (Ed. Kresta, J.E.), *ACS Symp. Ser.* **270** (1985), Chapter 10.
44. Ivin, K.J. and Saegusa, T. (eds.), *Ring-opening Polymerization*, Elsevier, London (1984).
45. McGrath, J.E. (ed.), Ring-opening Polymerization, *ACS Symp. Ser.* **286** (1985).
46. Reference 2, Volume 3, Chapters 31-37 and 45-53.
47. Webster, O.W., Hertler, W.R., Sogah, D.Y., Farnham, W.B. and RajanBabu, T.V. *J Amer. Chem. Soc.* **105**, 5706 (1983).
48. Price, C.C. and Carmelite, D.D. *J. Amer. Chem. Soc.* **88**, 4039 (1966).
49. Aida, T. and Inoue, S. *Macromolecules* **14**, 1166 (1981).
50. Asano, S., Aida, T. and Inoue, S. *J. Chem. Soc., Chem. Commun.* 1148 (1985).
51. Aida, T., Maekawa, Y., Asano, S. and Inoue, S. *Macromolecules* **21**, 1195 (1988).
52. Dale, J. and Daasvatn, K. *J. Chem. Soc., Chem. Commun.* 295 (1976).
53. Penczek, S., Kubisa, P., Matyjaszewski, K. and Szymanski, R. *Pure Appl. Chem.*, 140 (1984).
54. Penczek, S., Kubisa, P. and Szymanski, R. *Makromol. Chem., Macromol. Symp.* **3**, 203 (1986).
55. Miyamoto, M., Sawamoto, M. and Higashimura, T. *Macromolecules* **17**, 265 (1984).
56. Miyamoto, M., Sawamoto, M. and Higashimura, T. *Macromolecules* **18**, 123 (1985).
57. Miyamoto, M., Sawamoto, M. and Higashimura, T. *Macromolecules* **18**, 611 (1985).
58. Hashimoto, T., Ibuki, H., Sawamoto, M. and Higashimura, T. *J. Polym. Sci., Polym. Chem. Ed.* **26**, 3361 (1988) and references cited therein.
59. Maerov, S.R. *J. Polym. Sci.* **3**, 487 (1965).
60. Bellus, D., Manasek, Z., Hrdlovic, P. and Slama, P. *J. Polym. Sci., Part C* **16**, 267 (1967).
61. Chiellini, E. and Lenz, R.W. in Reference 2, Volume 5, Chapter 38.
62. Johnson, R.N., Farnham, A.G., Clendinning, R.A., Hale, W.F. and Merriam, C.N. *J. Polym. Sci., Polym. Chem. Ed.* **5**, 2375 (1967).

63. Rose, J.B. and Staniland, P.A. (ICI plc), European Patent 879 (1979).
64. Attwood, T.E., Dawson, P.C., Freeman, J.L., Hoy, L.R.J., Rose, J.B. and Staniland, P.A. *Polym. J.* **22**, 1096 (1981).
65. Tomalia, D.A., Baker, H., Dewald, J., Hall, M., Kallos, G., Martin, S., Roeck, J., Ryder, J. and Smith, P. *Polym. J.* **17**, 117 (1985).
66. Tomalia, D.A., Baker, H., Dewald, J., Hall, M., Kallos, G., Martin, S., Roeck, J., Ryder, J. and Smith, P. *Macromolecules* **19**, 2466 (1986).
67. Wilson, L.R. and Tomalia, D.A. *Polym. Prepr.*, **30**(1), 115 (1989).
68. Newkome, G.R., Yao, Z., Baker, G.R. and Gupta, V.K. *J. Org. Chem.* **50**, 2003 (1985).
69. Newkome, G.R., Yao, Z., Baker, G.R., Gupta, V.K., Russo, P.S. and Saunders, M.J. *J. Amer. Chem. Soc.* **108**, 849 (1986).
70. Tomalia, D.A., Hall, M. and Hedstrand, D.M. *J. Amer. Chem. Soc.* **109**, 1601 (1987).
71. Tomalia, D.A., Berry, V., Hall, M. and Hedstrand, D.M. *Macromolecules* **20**, 1164 (1987).
72. Naylor, A.M., Goddard III, W.A., Kiefer, G.E. and Tomalia, D.A. *J. Amer. Chem. Soc.* **111**, 2339 (1989).
73. Wallen, L.L. and Rohwedder, W.K. *Environ. Sci. Technol.* **8**, 576 (1974).
74. Collins, S.H. European Patent 111,231 (1982).
75. Doi, Y., Kunioca, M. and Soga, K. *Macromolecules*, **19**, 2860 (1986).
76. Ballard, D.G.H., Courtis, A., Shirley, I.M. and Taylor, S.C. *Macromolecules* **21**, 294 (1988).
77. McKean, D.R. and Stille, J.K. *Macromolecules* **20**, 1787 (1987).

2 Group transfer polymerizations

G.C. EASTMOND and O.W. WEBSTER

2.1 Introduction

2.1.1 *Background*

Free-radical polymerization can be used to produce high molecular weight polymers from the greatest variety of monomers. However, this mechanism leads to polymers with relatively broad molecular weight distributions and allows little or no control over polymer microstructure. Scientific and commercial advantages may be gained if control can be exercised over the molecular weight and molecular weight distribution of a polymer as it is formed (especially if the ratio $\bar{M}_w/\bar{M}_n$ can be reduced to about unity), if tacticity can be controlled and if inherent chain termination can be avoided; the latter facilitates the synthesis of block and graft copolymers.

Essential steps in all chain-growth polymerizations are initiation, propagation, termination and, possibly, chain transfer. The overall kinetics of the process and the molecular weight distribution of the product depend on the relative rates of these processes. The synthesis of polymers with narrow molecular weight distributions requires essentially rapid initiation, compared with propagation, and little or no inherent termination or chain transfer. Assuming polymerization is not conducted in the vicinity of the ceiling temperature of the monomer,[1] complete conversion of the monomer is achieved and the degree of polymerization of the polymer is determined by the ratio of initial monomer and initiator concentrations.[2]

The first real progress towards meeting the above aims was made by developments in anionic polymerizations of styrene initiated by sodium naphthalenide[3] and alkyllithiums.[4] The reactions produce 'living' polymers with narrow molecular weight distributions ($\bar{M}_w/\bar{M}_n \approx 1.02$) and controlled degrees of polymerization. (A 'living polymerization' has no inherent termination reaction.) Sodium naphthalenide initiates by electron transfer to monomer to form a radical ion which dimerizes and propagates as a difunctional species (1) ($\bar{P}_n = 2[M]_0/[I]_0$). Alkyllithiums initiate by nucleophilic addition to the monomer to give a monofunctional propagating species (2) ($\bar{P}_n = [M]_0/[I]_0$).

The distinction is important for the synthesis of terminally functional reagents and block copolymers since the former leads to difunctional species capable of chain extension at both ends. These polymerizations of styrene are virtually free of termination, can produce polymers with very narrow molecular weight distributions and are the systems against which other living polymerizations are judged.

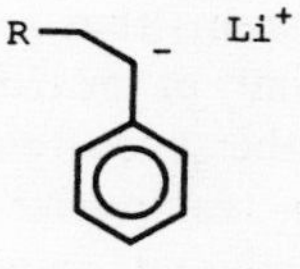

Whereas free-radical polymerization generally produces atactic (predominantly syndiotactic) polymers, anionic polymerization allows the synthesis of stereoregular polymers. Methacrylate polymerization in hydrocarbons initiated by butyl lithium leads to isotactic polymer[5] and in tetrahydrofuran (THF) to primarily syndiotactic polymer.[6]

Most cationic polymerizations are characterized by extensive chain transfer, but the synthesis of certain polymers with narrow molecular weight distributions is now possible. Polytetrahydrofuran with $\bar{M}_w/\bar{M}_n \approx 1.1$ can be prepared.[7] Vinyl ethers,[8-11] isobutylene and styrene can undergo living cationic polymerization.[12]

The synthesis of polymers with narrow molecular weight distributions by ionic techniques is successful with only a few monomers. Anionic polymerization of polar monomers is associated with extensive termination reactions. Methyl methacrylate (MMA) can only be polymerized to a product with a narrow molecular weight distribution at temperatures below about $-75\,°C$. Past reviews[13,14] have listed reactions in Scheme 2.1 as inherent termination processes. Reaction (a), involving addition to the carbonyl group and elimination of methoxide, may proceed readily if the active species, R^-, is the initiator. Evidence for this reaction involving the propagating anion is less

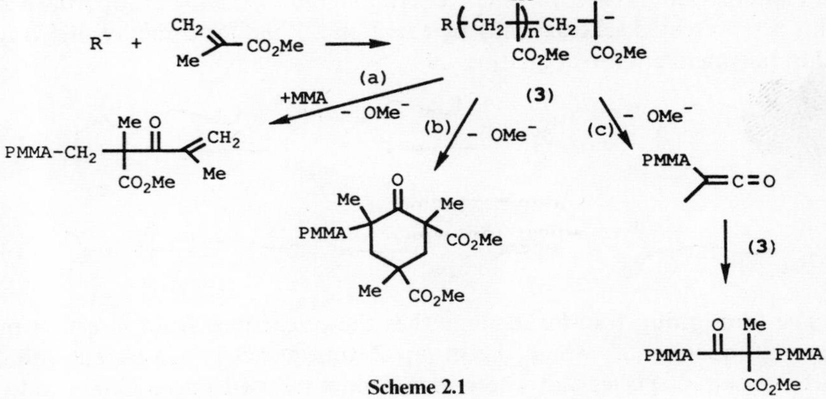

Scheme 2.1

clear. It now seems that reaction (b) is the predominant chain termination process, especially at the trimer stage[15] and that the methoxide ion eliminated moderates further occurrence of this reaction.[16]

Termination reactions broaden the molecular weight distribution of the polymer formed and prevent the synthesis of block copolymers and the effective formation of telechelic polymers by using a termination reaction of choice.

It was against this background that considerable interest was raised when DuPont announced the development of group transfer polymerization (GTP) of MMA as a living polymerization capable of producing polymers with narrow molecular weight distributions and block copolymers.[17-19]

2.1.2 *Terminology*

The term 'group transfer polymerization' (GTP) was introduced[17] to identify two new polymerization processes. In the original group transfer polymerization, based on silyl ketene acetals as the active species, a group from the terminal unit of the growing end of the chain is transferred to the incoming monomer during propagation as is depicted schematically in equation 2.1. For methyl methacrylate polymerization this may be written as in equation 2.2.

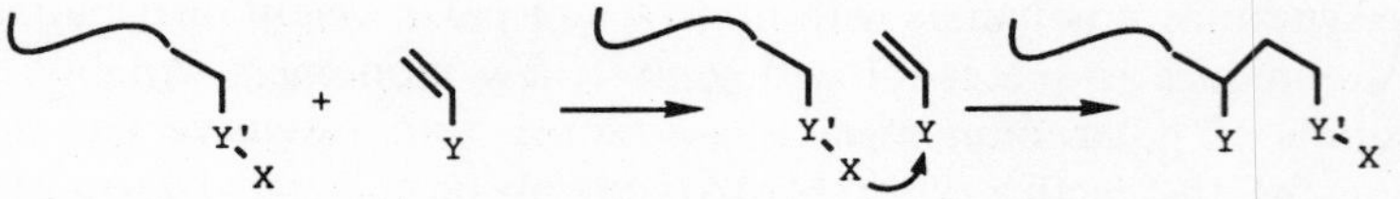

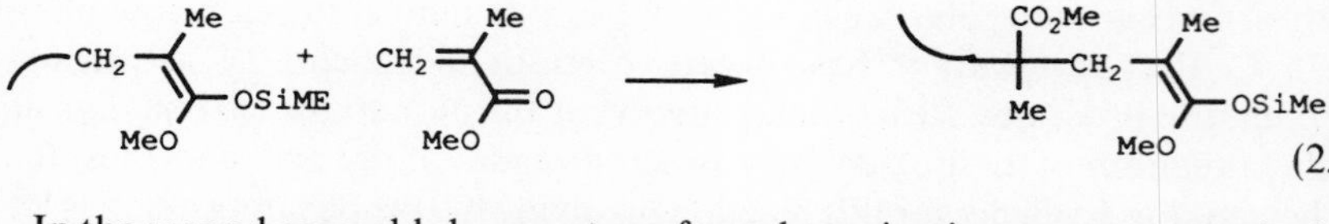

In the second case, aldol group transfer polymerization, a group attached to the incoming monomer is transferred to the terminal unit of the growing chain; all monomer units, except the active terminal unit, have the group attached. This is represented schematically by equation 2.3 and for trimethylsilyl vinyl ether polymerization by equation 2.4.

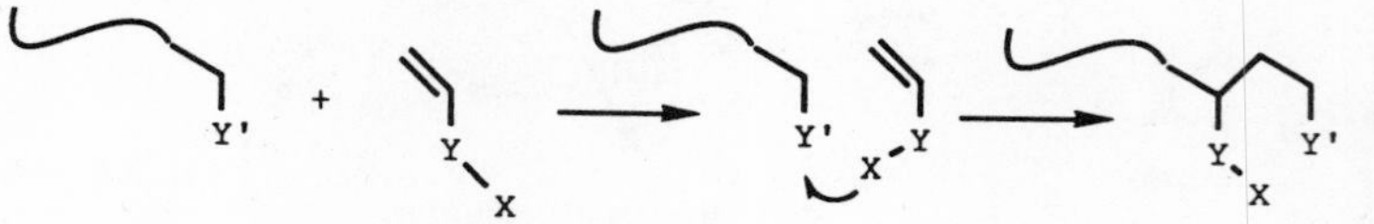

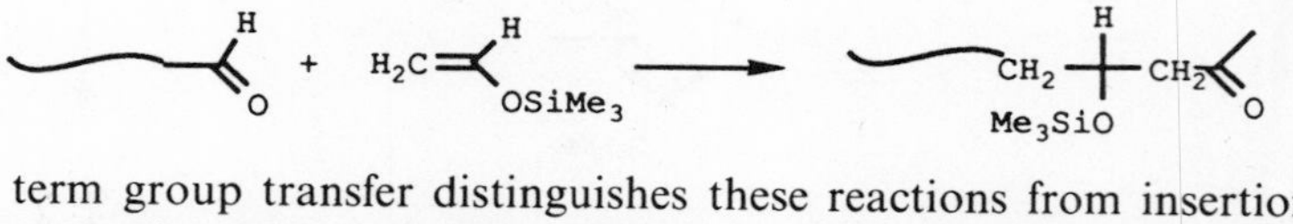

The term group transfer distinguishes these reactions from insertion reactions, equation 2.5, where the group Z transferred is associated with a carbon-centred species and where monomer is inserted into a C–Z bond.

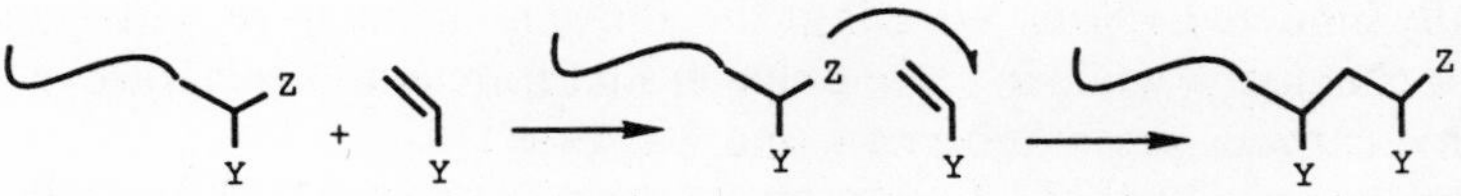

(2.5)

In view of the uncertainty of the exact nature of propagating species in ionic polymerizations, its variability with reaction conditions and the emergence of some newer polymerizations, a question arises as to how specific this terminology actually is. In anionic polymerizations of methacrylates in hydrocarbon solvents the negative charge is considered delocalized and the cation associated with the enolate anion; the negative charge is concentrated on the oxygen atom and the species are possibly associated in solution. Insertion of a monomer between the propagating species and the counter ion might be considered a group transfer process. Propagation in the polymerization of methacrylates and acrylates by metallo porphyrins is written as a group transfer reaction[20,21] as are newer anionic polymerizations initiated by metal-free carbanion salts.[22,23] A distinction which may be drawn between group transfer polymerization and, say, anionic polymerization is that in the former case the propagating species is a stable molecule which requires activation by a catalyst to participate in propagation.

Here we review the original group transfer polymerizations and consider the relationship between their mechanisms and other possibly related polymerizations later.

2.1.3 Nucleophilic addition

The main relevance of GTP is to the polymerization of acrylates and methacrylates. We therefore review some background chemistry which makes anionic polymerization of these monomers difficult but polymerization by GTP more straightforward.[24]

In nucleophilic addition to α,β-unsaturated carbonyls, equation 2.6, conjugate 1,4-addition (a), i.e. Michael addition, competes with 1,2-addition (b). On work up with acid, reaction (a) leads to overall addition to the C=C bond of the substrate and (b) leads to an unsaturated alcohol. 1,2-addition is favoured by reactive carbanions, including lithium and Grignard reagents, while 1,4-addition is favoured by resonance-stabilized carbanions. 1,2-Addition can be avoided either by making the nucleophile or the carbonyl

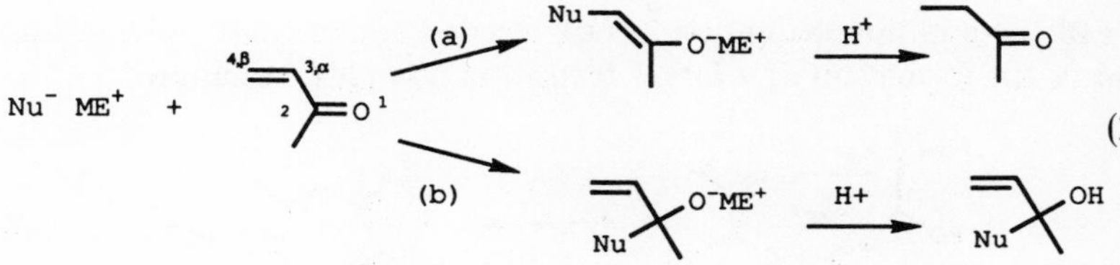

(2.6)

sterically hindered. (Note: We adopt the above nomenclature, widely used in organic chemistry, while in some polymer literature a nomenclature in which the vinyl carbons are numbered 1 and 2 is used.)

Immediate products of 1,4-addition are the enolate anions (**4, 5**) in which the charge is delocalized but resides primarily on the O-atom; the product of 1,2-addition is an alkoxide.

(4)   (5)

Kinetically, 1,2-addition is favoured, equation 2.7. However, if the attacking carbanion is sufficiently stabilized, because of the relative stabilities of it and the alkoxide ion, the attack will be reversible. Thermodynamically, 1,4-addition is favoured and the product (**6**), formed by the addition of the carbanion at the β-position, may build up.

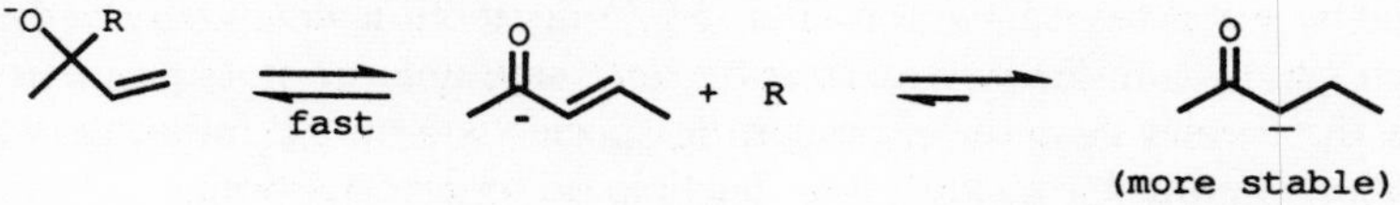

$$\text{(2.7)}$$

(more stable)
(6)

It is this competition between 1,2- and 1,4-addition, and the fates of the species formed, which gives rise to problems in the anionic polymerization of α,β-unsaturated carbonyls, including acrylates and methacrylates. In these cases the enolate ion is the ester enolate (**5**).

First, the carbanion of the initiating species should add at the β-position. Addition at the carbonyl can be avoided by choosing a sterically hindered initiator such as 1,1-diphenylhexyl lithium.[25]

Second, the ester enolate ion should undergo 1,4-addition to the monomer (electrophile), forming the new C—C bond at the C-atom of the enolate. This requirement is achieved if the cation (usually Li^+) associates with the enolate anion and the electronegative end of the incoming monomer; if the monomer is polarized and the oppositely polarized C-atom approaches in a 6-membered ring the result is 1,2-addition and the formation of alkoxide, equation 2.8.

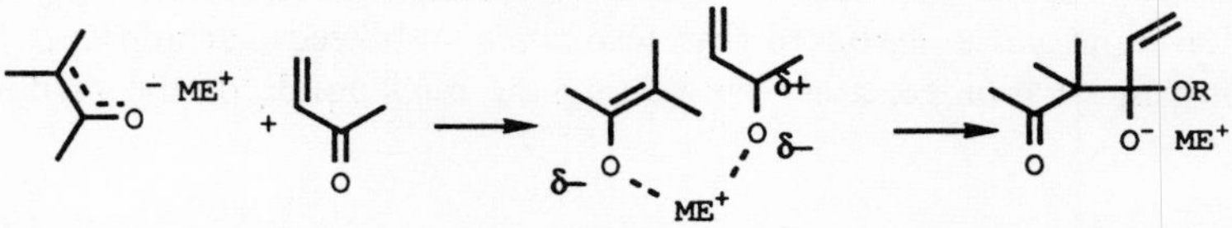

$$\text{(2.8)}$$

Overall 3,4-addition (equation 2.6(a)), needed for normal polymerization, requires the formation of a larger transition complex, equation 2.9.

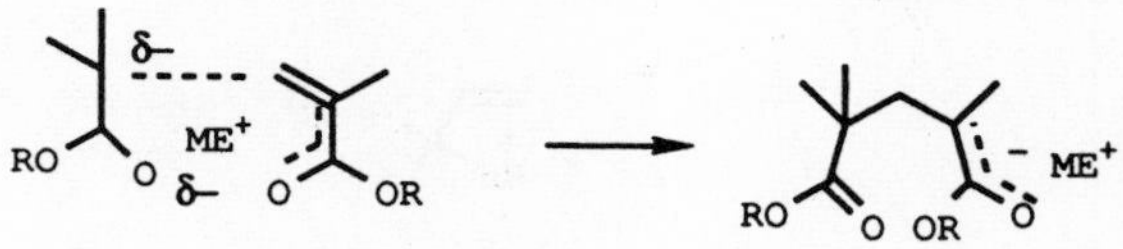

$$\text{(2.9)}$$

Polymerization requires repeated conjugate addition, enchaining monomeric residues through net 3,4-addition. Conditions are required where competition between different modes of addition is forced in one direction. Also, reaction conditions must not allow alkoxide elimination which can, in principle, follow either 1,2- or 1,4-addition to form unsaturated esters in the former case, (equation 2.10), and to intramolecular cyclization in the latter, (equation 2.11).

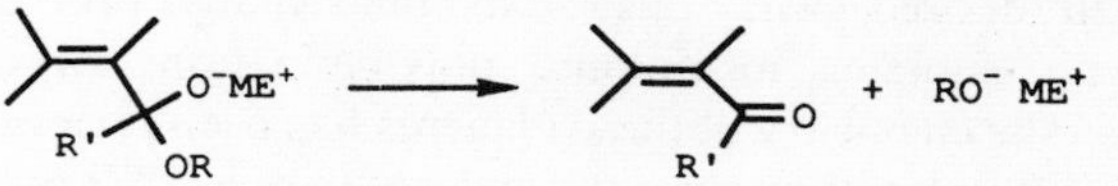

$$\underset{R'}{\diagup}\!\!=\!\!\underset{OR}{\diagdown}\!\!\overset{O^-ME^+}{} \longrightarrow \underset{R'}{\diagup}\!\!=\!\!\diagdown\!\!=\!\!O + RO^- ME^+ \qquad (2.10)$$

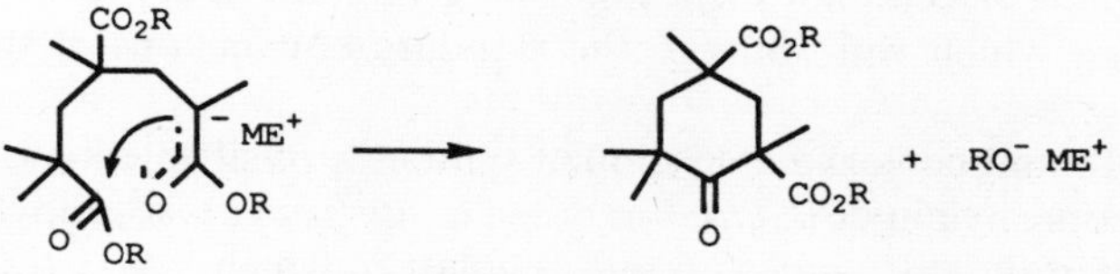

$$[\text{cyclic intermediate with } CO_2R, OR, O^-, ME^+] \longrightarrow [\text{cyclohexanone with two } CO_2R] + RO^- ME^+ \qquad (2.11)$$

It is now established that enolate anions are stable at $-78\,°C$ and undergo repeated addition in the desired way. At higher temperatures 1,2-addition and alkoxide elimination can occur and deviations from ideal polymerization behaviour become apparent. The temperature at which these deviations become significant depends on the substituent R in the ester group. Thus, anionic polymerization of methyl methacrylate to living polymers with a narrow molecular weight distribution can be performed successfully only below about $-70\,°C$ while t-butyl methacrylate can be polymerized at $+24\,°C$.[26]

This situation is further complicated in the case of acrylate polymerizations when the hydrogen attached to the α-carbon atom is relatively acidic and can be abstracted by the enolate anion.

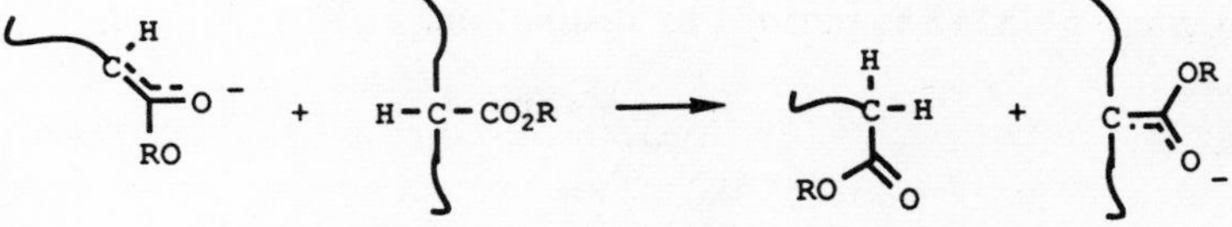

$$\underset{RO}{\diagup}\!\!C\!\!=\!\!O^- + H\!-\!C\!-\!CO_2R \longrightarrow \underset{RO}{\diagup}\!\!C(H)\!-\!H + C\!\!=\!\!\underset{O^-}{\overset{OR}{\diagup}} \qquad (2.12)$$

The possibility of overcoming these various difficulties and thereby enhancing control over the polymerization of acrylates and methacrylates, through the use of trapped enolates, created considerable interest in GTP.

2.1.4 Silyl ketene acetals

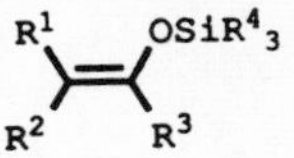

$$\underset{R^2}{\overset{R^1}{\diagdown}}\!\!=\!\!\underset{R^3}{\overset{OSiR^4{}_3}{\diagup}} \qquad\qquad \underset{R^2}{\overset{R^1}{\diagdown}}\!\!=\!\!\underset{OR^3}{\overset{OSiR^4{}_3}{\diagup}}$$

(7) 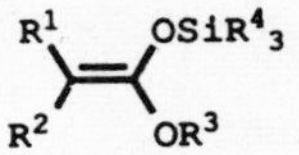 (8)

Silyl enol ethers (7) and silyl ketene acetals (8) came into prominence recently as versatile reagents which provide excellent regiochemical control over several organic reactions, including addition to α,β-unsaturated esters and ketones. These reagents are trapped enolates which, in the presence of suitable reagents (bases), are sources of enolates in reaction with electrophiles; they can also undergo reactions without generation of the enolate. Because of their regioselectivity, they are often used in preference to enolates. Although reactive, they are not unduly difficult to handle and the silicon moiety is readily removed when desired. They also overcome many problems generally associated with enolates in anionic polymerizations of acrylates and methacrylates. The chemistry of these reagents has been reviewed[27-34] and is pertinent to GTP technology since the end group of a GTP polymer is a silyl ketene acetal which will undergo the standard conversions of this class of reagent.

Ainsworth and coworkers developed syntheses of silylated enolates using lithium diisopropylamide (LDA) (prepared *in situ* from di-isopropylamine and butyl lithium) to create an enolate which is trapped by trimethylsilylchloride,[35,36] equation 2.13. Silyl ketyl acetals, the preferred

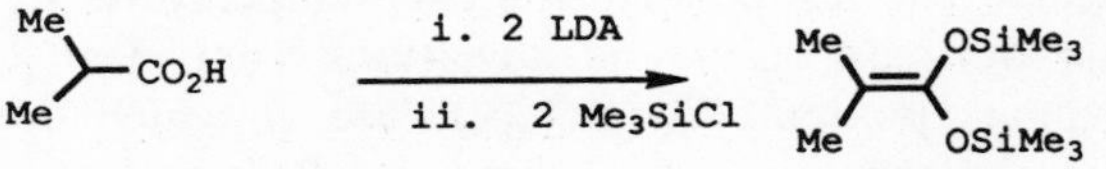

$$\underset{Me}{\overset{Me}{>}}\!\!-CO_2H \quad \xrightarrow[ii.\ 2\ Me_3SiCl]{i.\ 2\ LDA} \quad \underset{Me}{\overset{Me}{>}}\!\!=\!\!\underset{OSiMe_3}{\overset{OSiMe_3}{<}} \tag{2.13}$$

(9)

reagents as initiators for GTP, are readily prepared from appropriate esters.

For good molecular weight control in GTP, initiation must be fast compared with propagation. Thus, in the absence of any reason for using specially functionalized initiators, the usual and simplest initiator is 1-methoxy-1-(trimethylsiloxy)-2-methyl-propene (MTS) (**10**), the model compound for the propagating species in methyl methacrylate polymerization. Available commercially, MTS is readily prepared from methyl isobutyrate;[37,38] MTS is purified by distillation.

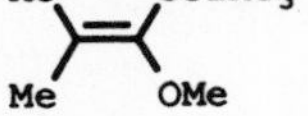

$$\underset{Me}{\overset{Me}{>}}\!\!=\!\!\underset{OMe}{\overset{OSiMe_3}{<}}$$

(10)

While the yields achieved in highly regioselective reactions are generally excellent for synthetic organic reaction, the desired yields of continuous additions in propagation to produce high molecular weight polymer must exceed 99%.

2.2 Features of group transfer polymerization

2.2.1 *General features*

The influences of several factors which affect group transfer polymerization are reflected in a few experimental parameters. It is useful to define the most common parameters in terms of the simplified mechanism in Scheme 2.2 where k_i is the rate coefficient for addition of the initiator, assumed to be a silyl ketene acetal, to monomer to generate a propagating species. $k_{p,n}$ is the rate coefficient for propagation by addition of monomer to a propagating chain having n main-chain units derived from the monomer. $k_{t,n}$ is a rate coefficient for any inherent termination process and k_{tx} is that for reaction with an added terminating agent X. k_f is a rate coefficient for transfer to substrate S; S^* initiates another propagating chain.

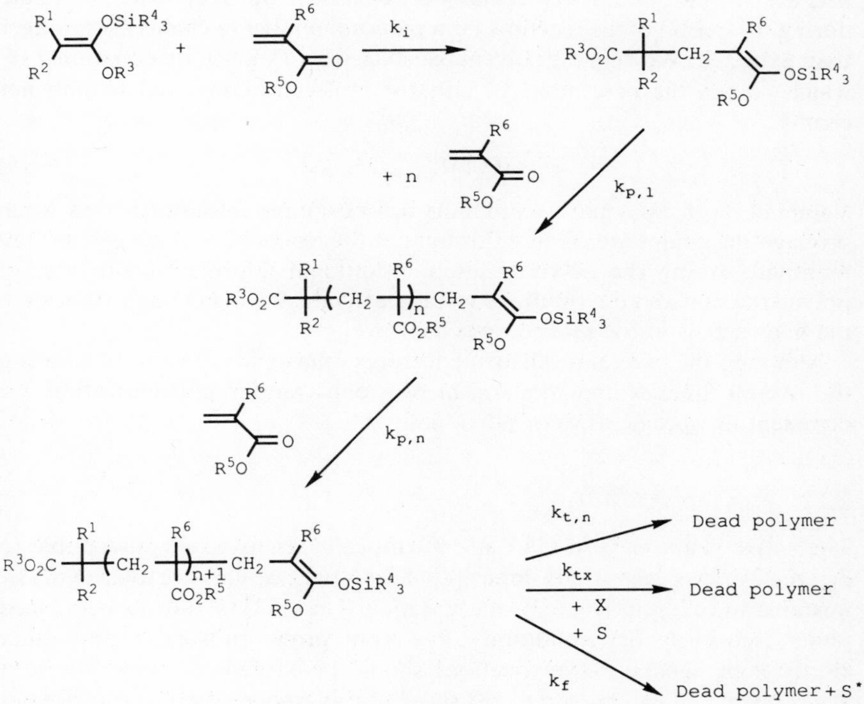

Scheme 2.2

This simplified scheme is incorrect in that reaction may not proceed in the absence of the catalyst which is left out of Scheme 2.2; Scheme 2.2, however, allows us to note that for a living polymerization ($k_{t,n} = 0$) in the absence of terminating agent ($[X] = 0$) and in the absence of chain transfer ($[S] = 0$ or $k_f = 0$). Assuming all initiator molecules initiate polymer chains, then, at the end of the polymerization, the number average degree of polymerization, P_n, and number average molecular weight $\bar{M}_n$ expected are given by

$$\bar{P}_n = [M]_0/[I]_0, \quad \bar{M}_{n,\text{calc}} = ([M]_0/[I]_0)M_0 \qquad (2.14)$$

where $[M]_0$, $[I]_0$ are initial concentrations of monomer and initiator and M_0

is the molecular weight of the monomer unit; strictly the molecular weight of the monomer unit incorporated from the initiator and any unit derived from the terminating agent should be added. The relations in 2.14 will hold even if $k_{t,n} \neq 0$, and all monomer is consumed prior to all chains being terminated. If, in addition, $k_i \geqslant k_p$ and $k_{t,n} = 0$, then a narrow molecular weight distribution ($\bar{M}_w/\bar{M}_n \approx 1$) can be expected.

Many GTPs do not give results in accordance with these expectations. In many cases the observed molecular weights ($\bar{M}_{n,exp}$) are greater than $\bar{M}_{n,calc}$ because not all initiator molecules start chains or not all chains propagate during the period of the reaction, i.e. a reduced number of chains grow longer than expected. We define a percentage efficiency (f), given by equation 2.15, which counts the percentage of initiator molecules converted to polymer chains.

$$f = 100(\bar{M}_{n,calc}/\bar{M}_{n,exp}) \qquad (2.15)$$

Values of D ($\bar{M}_w/\bar{M}_n$) may exceed unity if, for example, initiation is slow when propagating chains are created throughout the reaction, or some chains may terminate during the polymerization. Additional information on rates of polymerization and distributions would be invaluable in helping to elucidate the finer details of the reaction mechanisms.

Although the several mechanistic features inter-relate, prior to discussing the overall kinetics and mechanism of group transfer polymerization, we comment on specific aspects individually.

2.2.2 Reaction conditions

The active chain ends in GTP are intrinsically reactive and susceptible to destruction by adventitious impurities. Silyl ketene acetals are hydrolytically unstable and are sensitive to protic materials. Thus GTP should be performed under rigorously dry conditions, free from protic substances and other electrophilic agents. Polymerizations should be carried out under an inert atmosphere and can be performed satisfactorily under experimental regimes used for air-sensitive materials, employing syringe techniques.

Monomers may be purified by passing through neutral alumina (Brockmann No. 1). Any of several unreactive solvents may be employed although it is now recognized that acetonitrile, used to dissolve certain catalysts, may become silylated and induce complications.[39] On the other hand, the active species are not sensitive to some groups which might interfere with polymerization reactions in general. Advantage can be taken of these features to incorporate desired functionality into polymers synthesized by GTP (see Section 2.2.5).

GTP is not especially sensitive to temperature, reactions may be performed below, at or above ambient, up to $\sim 100\,°C$. Unwanted side reactions increase with increasing temperature. The polymerization is exothermic and onset of reaction may be detected by a temperature rise; the heat of polymerization can be removed by allowing low boiling solvents to reflux.

2.2.3 Initiators

This section is primarily concerned with initiators for the polymerization of α,β-unsaturated esters and for MMA in particular. For GTP, the most common, often preferred, initiator is 1-methoxy-1-(trimethylsiloxy)-2-methylprop-1-ene (MTS) (**10**), i.e. (**8**) with $R^1, R^2, R^3, R^4 = Me$, the model analogue of the propagating species.

Use of (**10**) for MMA polymerizations produces PMMA with a hydrogen derived from the initiator at one end. Termination with a proton source attaches a hydrogen atom at the other end, equation 2.16. We discuss later how special initiators give polymers with other end-groups.

$$\underset{(10)}{\overset{Me\ \ \ \ OMe}{\underset{H-H_2C\ \ \ \ OSiMe_3}{\diagup\!\!\!=\!\!\!\diagdown}}} + n\ CH_2\!=\!\!\underset{CO_2Me}{\overset{Me}{\diagdown}} \overset{Cat}{\longrightarrow} H\!\!-\!\!\!\left(\!CH_2\!-\!\!\underset{CO_2Me}{\overset{Me}{|}}\!\right)_{\!\!n}\!\!CH_2\!-\!\!\underset{OSiMe_3}{\overset{Me}{\diagup\!\!\!=\!\!\!\diagdown}}\!\!-\!OMe$$

$$\Big\downarrow H^+$$

$$H\!\!-\!\!\!\left(\!CH_2\!-\!\!\underset{CO_2Me}{\overset{Me}{|}}\!\right)_{\!\!n+1}\!\!\!H \qquad (2.16)$$

Effects of changing the groups on the silicon atom, as seen in comparisons between theoretical and experimental molecular weights and polydispersity, are not great until the substituents become bulky, i.e. (**8**) with $R^4 = Et$ gives equally good results as (**10**). Replacing one methyl on (**10**) with $C_{18}H_{37}$ reduces the initiator efficiency to 90% but increases the polydispersity ($D = 2.07$) significantly. However, replacing one methyl by t-butyl increases $\bar{M}_n$ of the polymer compared with the theoretical ($f = 11$) and increases the polydispersity ($D = 2.12$). Polymerizations with this initiator are very slow,[40] as if steric effects on Si affect k_p; k_i is expected to be affected comparably. If all rate coefficients were reduced equally similar results should be obtained but with lower rates of polymerization. Other effects must come into play, such as an effect on the catalyst (Section 2.2.4) or some form of chain termination (Section 2.2.6). The use of (**10**) with one phenyl group on the silicon has been mentioned.[40]

The related initiator (**11**) gave good agreement between observed and theoretical molecular weights and a value of $D = 1.10$, demonstrating that hydrogen attached to the silicon does not deleteriously affect the polymerization. In contrast, (**12**) ($R^1, R^3, R^4 = Me, R^2 = H$) exhibits a reduced efficiency ($\approx 70\%$) but D similar to that obtained with the preferred initiations (up to ≈ 1.4).[40] Possibly the catalyst used may have catalysed isomerization

(Section 2.2.6.1) of the O-silyl initiator to a less reactive C-silyl compound.[40] MTS isomerizes much more slowly than (12).

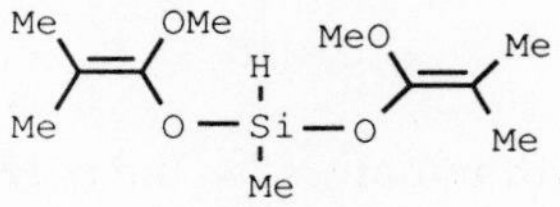

(11) 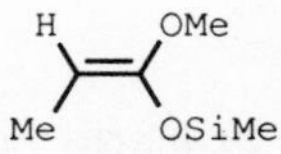   (12)

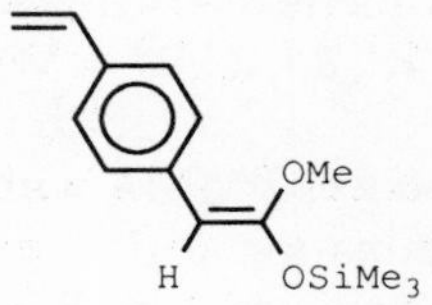

(13) 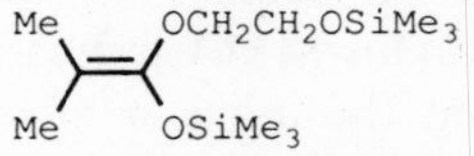   (14)

Asami et al.[41] used (13) which gave good results with $f = 83-100$ and $D \approx 1.1$ (results were somewhat irreproducible) to introduce a terminal, polymerizable styryl residue onto the chain. Changes in R^3 in (8) allows the introduction of end-group functionality into the polymers. Thus (14) provides a terminal group from which the trimethylsilyl group can be removed, with the aid of refluxing methanolic tetrabutylammonium fluoride[40,42] or with dilute methanolic hydrogen chloride at ambient temperatures,[40] to give PMMA with a terminal hydroxyl group (PMMA-OH),[43,44] equation 2.17. If

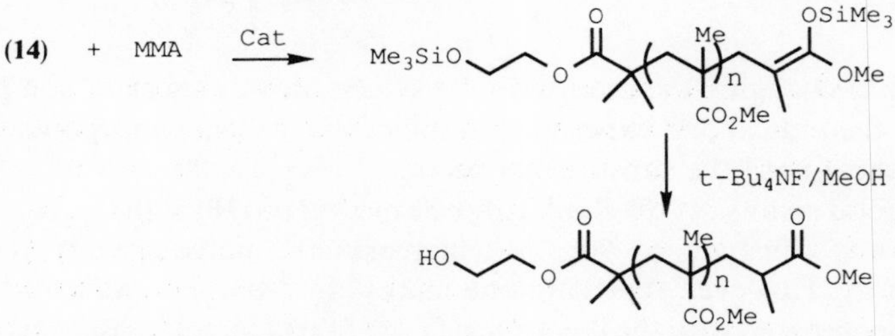

(2.17)

$R^3 = SiMe_3$ the initiator, with two silyloxy groups, is monofunctional and incorporates a capped carboxyl group at the initiator end of the chain. Deprotection by hydrolysis, as for the preparation of PMMA-OH, gives PMMA with a carboxy end group (PMMA-COOH),[17,44,45] equation 2.18. Deprotection may be carried out after chain termination (see termination).

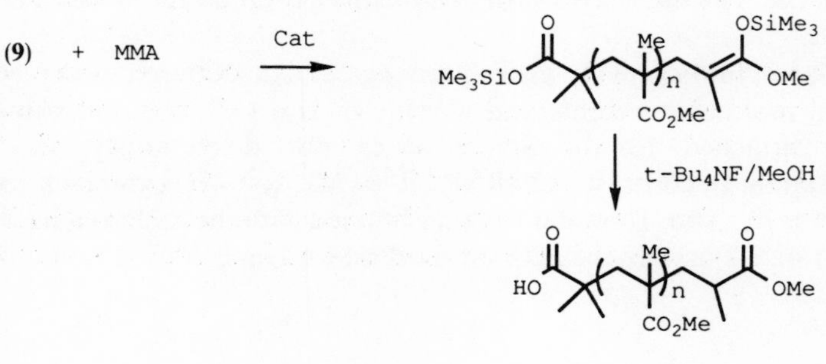

(2.18)

Some cyclic silyl ketene acetals also initiate GTP effectively,[40] e.g. **(15)**. Compound **(16)** was found to be a poor initiator and **(17)** to be somewhat better, using bifluoride as catalyst.[46]

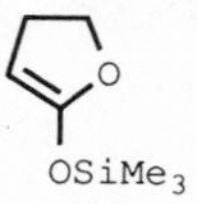

      (15) 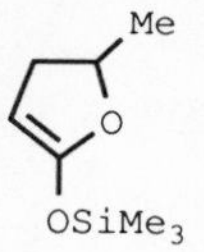   (16) 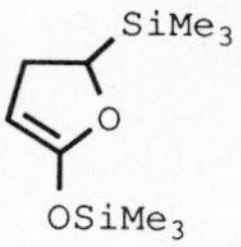   (17)

(11) is a potential difunctional initiator which might produce two propagating chains linked by Si, which would separate on termination. The corresponding compound in which the H on Si is replaced by a methyl group is claimed to be an effective difunctional initiator.[47] Other difunctional silyl ketene acetals have been used to produce propagating chains growing at both ends. Amongst these are **(18)**[46,48] and **(19)**.[48] These initiators produce PMMA with narrow molecular weight distribution but with $f \approx 60$.[48] Although it is likely that these initiators are difunctional, and such assumptions have been made,[48] propagation at both ends of the chain has not been established. **(20)** (as a mixture of isomers, 58% ZZ, 14% EZ, 28% EE) gave low conversions to polymer.[49]

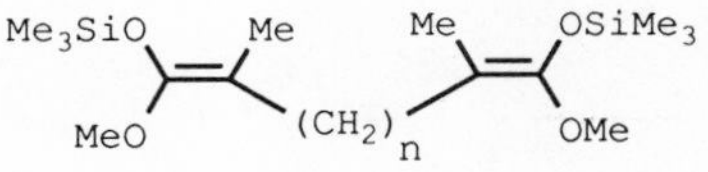

(18), n = 1
(19), n = 2

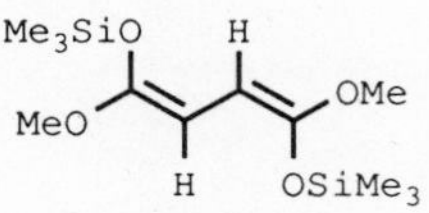

(20)

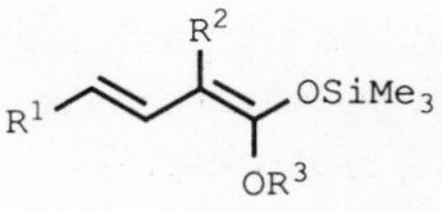

(21)

Silylpolyenolates of the type **(21)** have been examined[49,50] for initiation including

(22) $R^1, R^2 = H, R^3 = Et$; **(23)** $R^1 = H, R^2 = Me, R^3 = Et$;

(24) $R^1, R^2 = H, R^3 = OSiMe_3$; **(25)** $R^1 = H, R^2 = Me, R^3 = OSiMe$.

These initiators gave higher rates of polymerization of MMA than did MTS and good molecular weight control ($D = 1.12$–1.46) but $f = 67$–87. They provide opportunities for regioselective initiation. Thus **(22)** and **(24)** give **(26)** exclusively, while **(23)** and **(25)** give **(26)** and **(27)** in the ratio 2.5:1. The related initiator **(28)** gives **(29)** exclusively on initiation of MMA.[49,50] The related compound (30) initiates slowly and gives poor molecular weight control ($f = 25$–35, $D = 2.5$–2.7).[49,50]

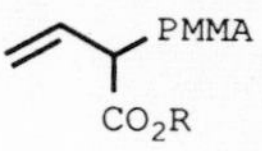

(26) CH$_2$=CH–CH(PMMA)–CO$_2$R

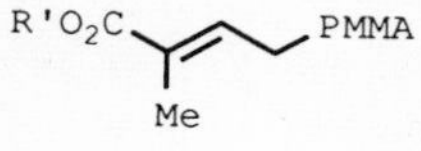

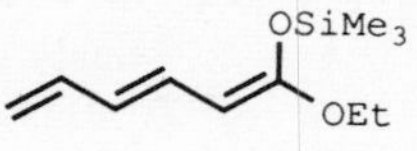

(27) R'O$_2$C–C(Me)=CH–CH$_2$–PMMA

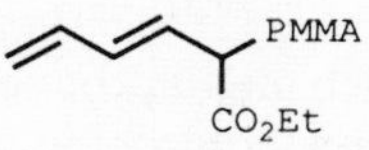

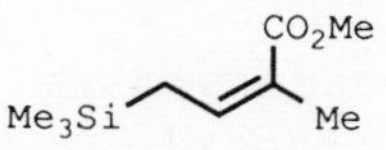

(28) CH$_2$=CH–CH=CH–C(OSiMe$_3$)(OEt)

(29) CH$_2$=CH–CH=CH–CH(PMMA)(CO$_2$Et)

(30) Me$_3$Si–CH$_2$–C(Me)=CH–CO$_2$Me

Other silicon reagents initiate GTP including Me$_3$SiCN,[17,40,51,52] Me$_3$SiSMe,[18,19,40,53,54,55,56] Me$_3$SiCH$_2$COOEt ($f = 18$),[40] Me$_3$SiCH$_2$COO-t-Bu ($f = 15$),[40] Me$_3$Si(CN)Me$_2$ ($f = 21$),[40] Me$_3$SiSPh,[40] Me$_3$SiCR$_2$CN[40] and R$_2$POSiMe$_3$.[40,57] Trimethylsilylcyanide may be generated in situ from R$_4$N$^{\oplus}$CN$^{\ominus}$ and Me$_3$SiCl.[58] In general these initiators exhibit low efficiency and give polymers with high polydispersity. They act by generating ketene acetals on addition to monomer, equation 2.19;[51] the product of initiation may undergo isomerization (Section 2.2.6.1).[51]

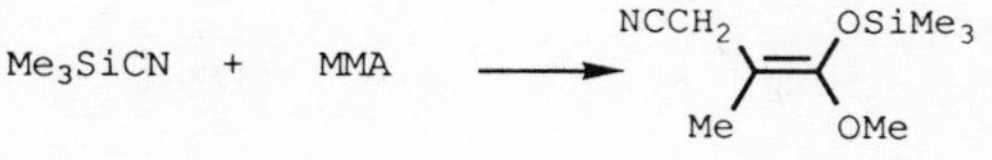

$$\text{Me}_3\text{SiCN} + \text{MMA} \longrightarrow \underset{(31)}{\text{NCCH}_2\text{C(Me)}=\text{C(OSiMe}_3)(\text{OMe})} \quad (2.19)$$

However, trimethylsilylcyanide (TMSCN) does not react with MMA in the absence of a catalyst such as tetramethylammonium cyanide, which acts as a source of CN$^{\ominus}$, or a source of other suitable anion such as F$^{\ominus}$. These systems exhibit induction periods during which (31) and oligomers from slowly, and TMSCN is slowly depleted; subsequently polymerization accelerates.[45] TMSCN possibly sequesters the catalyst anion and propagation can only proceed effectively when the TMSCN is depleted,[45] as in Scheme 2.3. Most polymerizations catalysed by TMSCN have been carried out in acetonitrile as solvent and this itself leads to further complications[39,51,59,60] (see later).

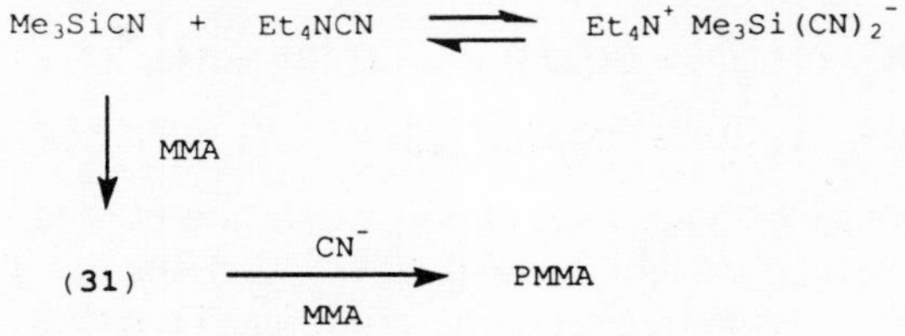

$$\text{Me}_3\text{SiCN} + \text{Et}_4\text{NCN} \rightleftharpoons \text{Et}_4\text{N}^+ \text{Me}_3\text{Si(CN)}_2^-$$

$$\downarrow \text{MMA}$$

$$(31) \xrightarrow[\text{MMA}]{\text{CN}^-} \text{PMMA}$$

Scheme 2.3

α-Trimethylsilyl, -stannyl and -germyl esters will initiate GTP but control over molecular weight and its distribution is reduced.[58] It has been suggested that the primary step is rearrangement to ketene acetals, equation 2.20, and that the slow rate of this process, and consequent slow initiation, is responsible for the broad molecular weight distributions.[40]

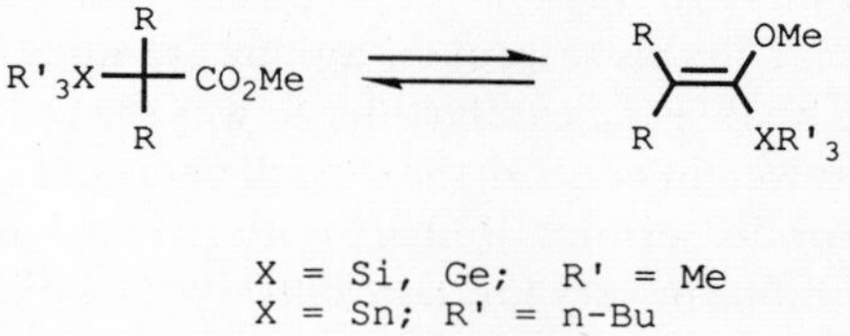

$$R'_3X\underset{R}{\overset{R}{-\!\!\!\!+\!\!\!\!-}}CO_2Me \rightleftharpoons \underset{R}{\overset{R}{>\!\!=\!\!<}}\overset{OMe}{\underset{XR'_3}{}} \qquad (2.20)$$

X = Si, Ge; R' = Me
X = Sn; R' = n-Bu

Titanium enolate (**32**) has been reported to initiate GTP at low temperatures.[52]

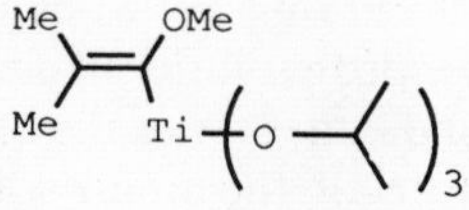

$$\underset{Me}{\overset{Me}{>\!\!=\!\!<}}\overset{OMe}{\underset{Ti(O-\langle\ \rangle)_3}{}}$$

(**32**)

2.2.4 Catalysts

Silyl ketene acetals are unreactive in GTP in the absence of a catalyst. Early literature on GTP referred to anionic (fluoride and bifluoride) and Lewis acid catalysts. Fluoride ion sources catalyse nucleophilic reactions of organosilanes,[29] including controlled additions to α,β-unsaturated esters and ketones; they are used in concentrations comparable to the silyl ketene acetal. Fluorides catalyse GTP but do not give controlled polymerizations. Polymerizations are less living than with preferred catalysts and polydispersities of the polymers are high. The catalysts interact with Si to form an Si–F bond which is strong compared to the Si–O bond which is broken to liberate an enolate. Fluoride catalysts have a role in termination reactions to produce telechelics (Section 6).

GTP introduced bifluorides (e.g. potassium bifluoride[58] and especially tris(dimethylamino)sulphonium (TAS) bifluoride (TASHF$_2$)) as catalysts for such processes;[17] TASMe$_3$SiF$_2$, azides and cyanides are also effective.[17,58] Although they are weaker catalysts for nucleophilic additions, these materials are better for GTP. They are considered to coordinate with the Si of the silyl ketene acetal to expand its valency, weaken the Si–O bond, change electron densities and facilitate addition. TASHF$_2$ is preferred and very low concentrations (as low as 0.01% of [initiator]) are said to be the optimum.[40,51] Polymerizations so catalysed are less prone to termination and the polymers have lower polydispersities. In spite of the sensitivity of GTP to protic reagents, tetrabutylammonium fluoride hydrate is a good catalyst;[40] the number of moles of water of hydration is small compared to the number of moles of initiator.

Later, catalysis by oxyanions and bioxyanions was announced.[61,62] Suitable anions are carboxylates (e.g. acetates and benzoates), phenolates, sulphinate, phosphinate and perfluoroalkoxide, biacetate and bibenzoates. In general, these catalysts are weaker but give better molecular weight control and data indicate that, in this respect, bi-anions are superior to mono-anions.

It seems that the weaker the catalyst the slower the polymerization but the better the molecular weight control; the overall value of k_p and the probability of termination (see under termination) are both reduced. Bi-anions may act as a source of mono-anions in very low concentration by dissociating according to equation 2.21.

$$(RCO_2)_2^{\ominus}H \rightleftharpoons RCO_2^{\ominus} + RCO_2H \tag{2.21}$$

There are claims that additions of so-called livingness enhancement agents[63,64] further reduce the probabilities of chain termination. Examples of such agents include the O-silylated ester of the oxyanion used as catalyst. It is also reported that the nature of the non-coordinating counterion of the catalyst influences the activity of the catalyst;[61] tetraalkylammonium and TAS cations are said to give the best results.

Clearly the influence of the catalyst is complex and subtle in determining the nature and concentration of true active chain ends, possibly through a series of equilibria.

Another factor which influences catalyst choice is solubility in polymerization media. Thus, potassium bifluoride is insoluble and the potential concentration of active species is unknown; crown ethers solubilize KHF_2 and produce higher rates of polymerization.[46,65] $TASHF_2$ has low solubility in THF and is usually added as a solution in acetonitrile; acetonitrile is associated with kinetic problems (Section 2.3.1.6). Other catalysts, such as tetrabutylammonium salts, are soluble in THF.

Alternatives to anion catalysts are Lewis acids such as zinc salts, dialkylaluminium chlorides and oxides.[40] Zinc salts are the preferred catalysts for acrylate polymerization and are used in 'high' concentrations, usually about 10 mol% based on monomer; aluminium catalysts are used at about 10 mol% based on initiator. These catalysts presumably coordinate with, and activate, the monomer.

Clay montmorillonite (sheet silicate), ion-exchanged with aluminium, catalyses Michael additions of silyl ketene acetals to α,β-unsaturated esters and polymeric by-products were reported for addition to acrylates.[66]

2.2.5 Monomers

The original patent literature on GTP[18] identifies several monomers. The process is most applicable to polymerizations of α,β-unsaturated esters, especially methacrylates and acrylates. Methacrylates are the preferred monomers and can be polymerized to polymers with narrow molecular weight distribution at temperatures up to 100 °C, although they are not totally free of complications (see termination). Methacrylates with bulky ester groups polymerize more slowly than does MMA. A methacrylate with a mesogenic unit, (6-[4-(4-methoxyphenoxycarbonyl)phenoxy]hexyl methacrylate), has been polymerized by GTP and gives a polymer with a nematic liquid crystalline mesophase.[67]

Acrylates, more reactive than methacrylates, have greater attendant problems. They generally lead to polymers with lower molecular weights than theory and with broader molecular weight distributions. To some extent these problems can be overcome using Lewis acid catalysts. Acrylates polymerize much more rapidly than do methacrylates and in a mixture of the two the acrylate polymerizes virtually exclusively (Section 2.4). Polymerization of acryloyloxyethyl methylacrylate (**33**) leads to a polyacrylate with pendant methacrylate residues,[69,70] equation 2.22).

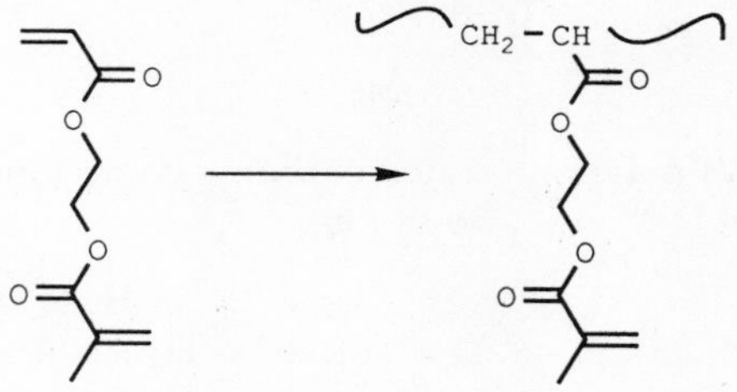

$$ \tag{2.22} $$

(**33**)

Functional groups susceptible to reaction in many polymerization media are inert in GTP. This can lead to polymers not normally accessible. Vinyl groups in hydrocarbon residues are inert and allyl and sorbyl residues in the ester groups of methacrylates remain in the polymer.[19] Styryl residues are unaffected.[68]

Glycidyl methacrylate (GMA) features in many formulations[19] and, if polymerized below 0 °C, provides pendant epoxy groups in the polymer. The epoxy groups may then be reacted to give functional polymers,[71] Scheme 2.4.

Polymers with pendant carboxyl and hydroxyl groups can be prepared if these functions in the monomer are protected. Thus, hydroxyethyl methacrylate units can be introduced using (**34**) as monomer.[19]

Monomers containing hydroxyl groups which form strong intramolecular hydrogen bonds may be polymerized without protection of the hydroxyl. Thus, ultraviolet stabilizers, 4-methacryloxy-2-hydroxybenzophenone and 2(2-hydroxy-4-methacryloxyphenyl)2H-benzotriazole, have been incorporated into PMMA by GTP.[72]

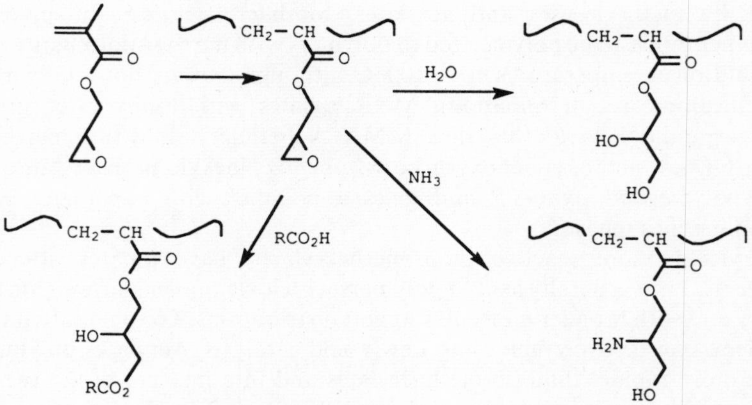

Scheme 2.4

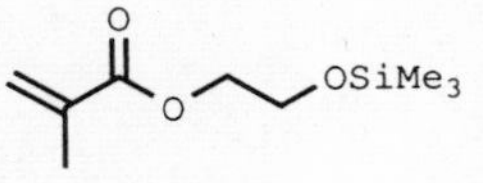

(34)

Polymerizations of polyunsaturated carbonyl compounds of the type (35) have been described,[49,50] and also of (36).

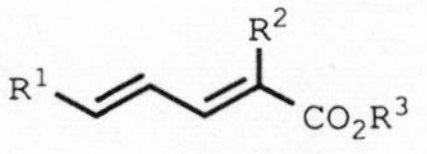

(a) $R^1, R^2 = H$; $R^3 = Me$
(b) $R^1 = H$; $R^2, R^3 = Me$
(c) $R^1 = H$; $R^2 = Me$; $R^3 = Et$
(d) $R^1 = Me$; $R^2 = H$; $R^3 = Et$
(e) $R^1 = Etox$; $R^2 = H$; $R^3 = Et$

Polymerization of (35a–e) gives polymers with one double bond per residue in the main chain, viz. (37), and (36) polymerizes to (38).

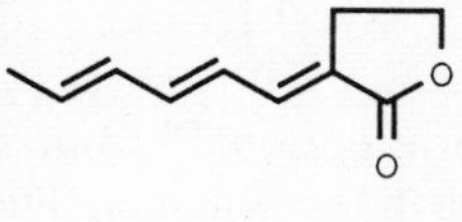

(36) 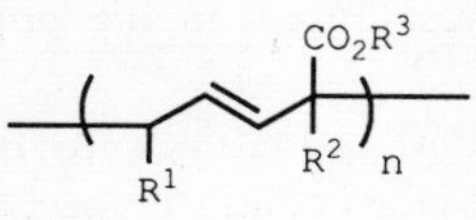   (37) 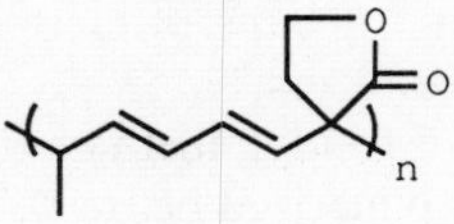   (38)

Polymerization of (35a) initiated by the analogue of the propagating species, catalysed by tetrabutylammonium acetate, proceeds cleanly. Related initiators give poorer molecular weight control, probably as a result of slow initiation compared with propagation. Polymerizations of (35b) and (35c) gave longer-lived polymers than (35a), presumably because they are analogous to methacrylates rather than acrylates. Polymerization of (36) initiated by MTS gave much higher $\bar{M}_w$ than expected, indicating slow initiation.

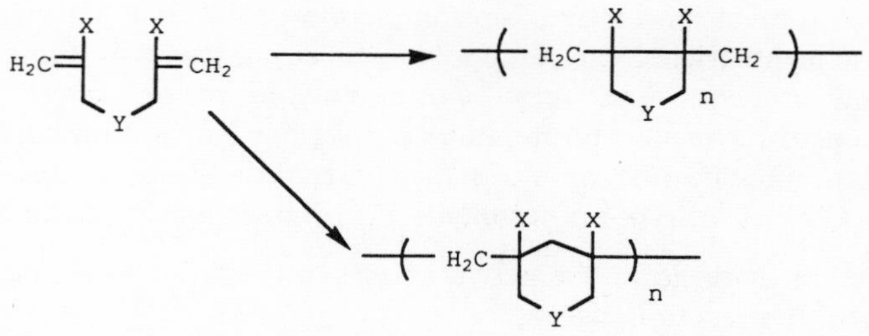

GROUP TRANSFER POLYMERIZATIONS

Scheme 2.5

Cyclopolymerization, which involves intramolecular cyclization accompanying polymerization through both vinyl groups, occurs in the free-radical polymerization of certain types of non-conjugated dienes, Scheme 2.5, to form 5- and/or 6-membered ring in-chain structures.[73] Kozakiewicz and Kurose found that polymerization of N-phenyldimethacrylamide, catalysed by TASF in the presence of acetonitrile, was slow, gave only 55% conversion of monomer and resulted in low molecular weight products. Polymerization resulted almost exclusively in formation of 6-membered ring units.

2.2.5.1 Ring-opening

Quirk and Bidinger[74] reported the ring-opening polymerization of propylene sulphide initiated by trimethylsilyl 2-trimethylsiloxyethyl sulphide; $TASHF_2$ was the catalyst and the solvent used was THF. Polymerization proceeded rapidly to low molecular weight polymer with $D = 1.12$ and $\bar{M}_{n,exp}$ as calculated by theory. Addition of MMA after polymerization of the propylene sulphide was complete, produced a further exotherm and the product showed the development of a new peak in the gel permeation chromatogram of the product; these data were taken as evidence for both the living nature of the propylene sulphide polymerization and the formation of a block copolymer. NMR spectroscopy showed the presence of two trimethylsilyl residues from the initiator in the product. The product formed by addition of allyl bromide and Bu_4NF at the end of the propylene sulphide polymerization showed the presence of an alkyl group in the polymer. It is presumed that the polymerization of propylene sulphide proceeds by GTP with the overall result as in equation 2.23.

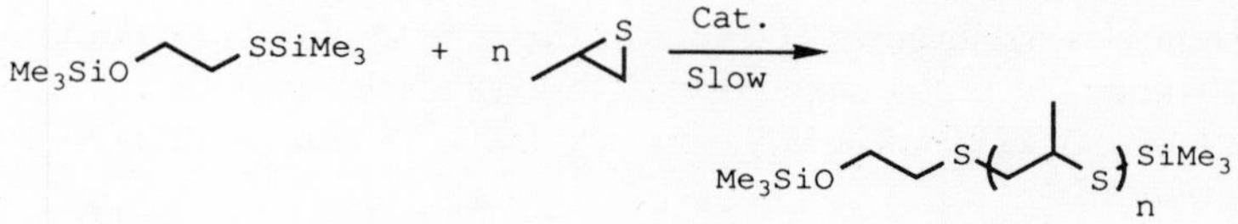

$$\text{Me}_3\text{SiO}\!-\!\!\sim\!\!-\text{SSiMe}_3 \; + \; n \; \overset{S}{\triangle} \quad \xrightarrow[\text{Slow}]{\text{Cat.}} \quad \text{Me}_3\text{SiO}\!-\!\!\sim\!\!-\text{S}\!\!-\!\!(\text{\textasciitilde})_n\!\!-\!\!\text{S}\!-\!\text{SiMe}_3 \tag{2.23}$$

2.2.6 Termination

Living polymerization implies an absence of inherent termination for propagating chains. Such reactions provide opportunities for addition of a second aliquot of monomer after completion of the first polymerization, a means of synthesizing block copolymers, or addition of some chosen terminating reagent. The term living polymerization often requires qualification to mean that termination is neligible on the time scale of the polymerization but may be significant on a longer time scale; this qualification applies to GTP. Two types of termination processes require consideration:

1. termination intrinsic to the active species and other species necessarily present in the reaction medium;
2. termination through adventitious impurities or added terminating agents, possibly to give terminal groups of choice.

2.2.6.1 *Inherent termination* GTP is not devoid of inherent termination. Two possible reactions must be considered: isomerization and backbiting.

Isomerization Silyl ketene acetals, in the presence of GTP catalysts, can isomerize to C-silylated esters, equation 2.24. For example, MTS (**10**) (R = Me in (**39**)) isomerizes to methyl α-(tri-methylsilyl)isobutyrate. Presumably the same reaction occurs with propagating chains (R = PMMA).

$$\underset{(39)}{\underset{R\quad\ OSiMe_3}{\overset{Me\quad\ OMe}{\diagup\!\!\!\diagdown}}} \longrightarrow R\underset{Me}{\overset{SiMe_3}{-\!\!\!\!\!\mid\!\!\!\!\!-}}CO_2Me \qquad (2.24)$$

α-silylated esters, e.g. ethyl (trimethylsilyl)acetate, will initiate GTP of MMA. Observed molecular weights are much greater than those calculated for rapid initiation and polydispersities are large. From such data it was concluded that initiation is much slower than propagation.[40] Because isomerization is reversible, O-to-C isomerization is not a true termination reaction. However, it effectively removes isomerized species from the pool of active chains; in the terminology of free-radical polymerization it represents retardation.

Some kinetic data on O-to-C and C-to-O isomerization are available.[40] For MTS, equilibrium corresponds to 76% O-silylated species. With 1% TASHF$_2$ catalyst, 5% O-to-C isomerization occurred in 133 min and 3% C-to-O isomerization in 116 min; isomerization gradually 'deactivated' the catalyst. With t-butylammonium m-chlorobenzoate as catalyst, starting with MTS, equilibrium was achieved in 800 min; in the reverse direction equilibration took 2000 min. It was concluded that O-to-C isomerizations do not compete with initiation of GTP of MMA catalysed by MTS and, possibly, not with propagation.[40]

Backbiting The inherent termination reaction in GTP is a backbiting process analogous to reaction (b) in Scheme 2.1, which can be represented by equation 2.25. This reaction, involving three monomer units, becomes prevalent for chains longer than dimer and is most serious for the trimer.

Details of this reaction are discussed in Section 2.3. Here we make some general comments.

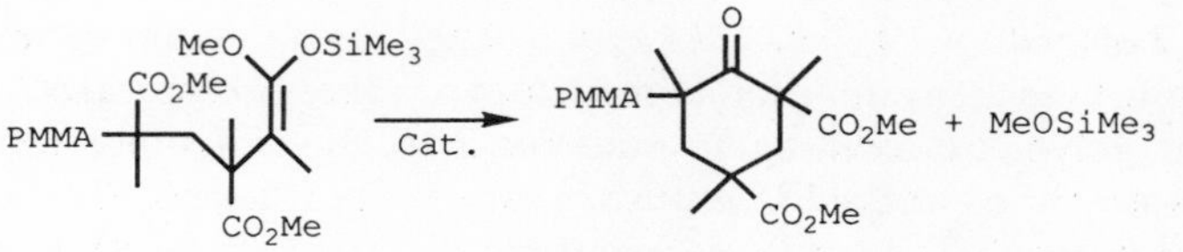

$$PMMA\text{—}\underset{CO_2Me}{\overset{MeO\ \ \ OSiMe_3}{\underset{|}{\overset{|}{C}}=\underset{|}{\overset{CO_2Me}{C}}}}\xrightarrow{Cat.} PMMA\text{—}\underset{CO_2Me}{\overset{O}{\underset{|}{\overset{||}{C}}}}\text{—}CO_2Me + MeOSiMe_3 \quad (2.25)$$

Neglecting any chain-length dependence of the rate coefficient, k_t, for termination, consider competition between propagation (rate coefficient, k_p) and termination according to Scheme 2.6.

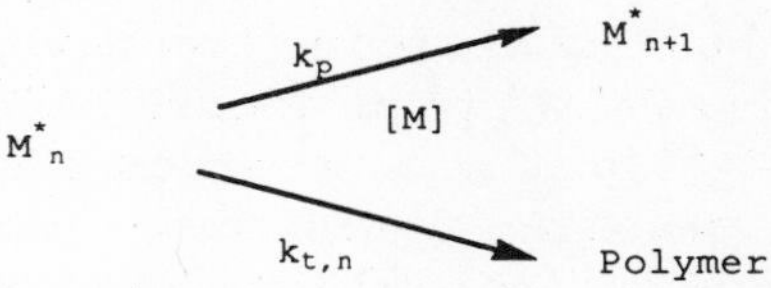

$$M^*_n \underset{k_{t,n}}{\overset{k_p\ [M]}{\rightrightarrows}} \begin{array}{l} M^*_{n+1} \\ \\ Polymer \end{array}$$

Scheme 2.6

The probability of propagation, p, is given by equation 2.26. To form high molecular weight polymer, p must approximate to unity. p tends to unity when either $k_t = 0$ or $[M] \gg k_{t,n}/k_p$. In GTP of MMA at 25 °C, k_p is an order of magnitude greater than k_t.

$$p = \frac{k_p[M]}{k_p[M] + k_{t,n}} = \frac{1}{1 + \dfrac{k_{t,n}}{k_p[M]}} \quad (2.26)$$

In the synthesis of low molecular weight PMMA and other methacrylate polymers, termination is not serious and high conversions to polymers with low polydispersities ($D \approx 1$) are achieved. However, as $[M]$ diminishes, backbiting takes place at a significant rate. Termination then has serious consequences in the syntheses of block copolymers and terminally functionalized polymers. In these cases polymerizations should not reach monomer-starved conditions before addition of the next reagent. In block copolymer synthesis addition of a second monomer prior to consumption of the first will lead to some tapering of the blocks, whilst complete polymerization of the first block will lead to some loss of efficiency in converting the first polymer to a block copolymer. In producing terminally functionalized polymers, backbiting prior to addition of the terminating reagent will reduce the functionality of the resulting polymer.

To minimize such effects, it is necessary to operate the reactions under conditions which reduce k_t relative to k_p, i.e. to keep the reaction temperature low. Relative values of k_t and k_p are also influenced by the catalyst and, possibly, by other added reagents (see Section 2.2.4). For example, bi-oxyanion catalysts, although giving rise to slower polymerizations, have lower probabilities of termination (see Section 2.3).

2.2.6.2 *Termination by added agent* The most common means of terminating GTP is by addition of methanol or other protic reagent to give a saturated polymer according to equation 2.16. Several other termination reactions described in the literature are used to form terminally reactive, or otherwise functionalized, polymers and are discussed under Section 2.5.

2.2.7 *Polymer microstructures*

Tacticities of PMMAs prepared by GTP have been determined.[40,69,75,76] Tacticity is influenced by the catalyst and polymerization temperature and not by the solvent used. For PMMA prepared using $TASHF_2$ as catalyst, data show systematic trends.[40] In all cases the content of isotactic triads is low ($<10\%$) while the syndiotactic content varies linearly with synthesis temperature from about 80% at $-90\,°C$ to about 50% at $60\,°C$. It was calculated that the differences in activation enthalpy ($\Delta\Delta H^{\ddagger}$) and entropy ($\Delta\Delta S^{\ddagger}$) between m and r diad formation are $4.2\,kJ\,mol^{-1}$ (favouring r placement) and 4.62 eu.[40] These values are comparable to those obtained for free-radical polymerization of MMA ($4.5\,kJ\,mol^{-1}$ and 4.16 eu).[77] Similar results were reported by Müller and Stickler[75] and by Wei and Wnek.[76] The triad populations agree closely with Bernoullian statistics. With bifluoride catalyst it was found that a polymethacrylate with bulky mesogenic ester groups gave a less syndiotactic polymer.[67] When a Lewis acid catalyst was employed the content of syndiotactic triads was twice that of heterotactic triads and independent of temperature.[69] These data imply that anionic and Lewis acid catalysts involve different transition states. Wei and Wnek also found poly(t-butyl methacrylate) prepared by GTP at $20\,°C$ to be less syndiotactic (38%) ($\Delta\Delta H^{\ddagger} \approx 1.4\,kJ\,mol^{-1}$ and $\Delta\Delta S^{\ddagger} \approx 0.3\,eu$) than PMMA (51%) and poly(trimethylsilyl methacrylate) to be more syndiotactic (66%).[76] The tacticity of PMA prepared using anionic catalysts is reported to be virtually independent of temperature, with approximately 20% isotactic and 31% syndiotactic triads.[76]

Brittain *et al.* found that in the isomerization of (**40**), equation (2.27), the first adduct of MTS to MMA, E is more stable than Z and isomerization is catalysed by HF_2^-, carboxylate anions and mercuric iodide.[78]

(Z-) ⇌ (E-) (2.27)

(**40**)

Hertler et al.[49] reported on the structure of polymers prepared from several polyunsaturated monomers.

By analogy with previous assignments made for NMR spectra of polymers prepared from α-methylene-γ-butyrolactone,[79] polymer prepared by GTP contained more mm sequences than the polymer prepared by free-radical polymerization.[40]

2.3 Kinetics and mechanism

2.3.1 *Polymerization of methacrylates*

Several studies provide an insight into the mechanism of GTP and the major steps of initiation, propagation, termination and chain transfer. Because on the time scale required to reach high monomer conversions polymerizations can be considered living, information on the individual steps has been obtained in separate studies.

2.3.1.1 Nature of propagating species The active species in GTP, capable of Michael addition to monomer, involves interaction of the nucleophilic catalyst with the silyl ketene acetal. In synthetic organic chemistry, fluoride catalysts are used in stoichiometric equivalence with silyl ketene acetals or silyl enol ethers. Results are consistent with the formation of a strong Si—F bond and abstraction of the trialkylsilyl group to form an enolate ion; reaction then occurs with the enolate. Under certain conditions Michael additions are reversible.[80]

Because, in GTP, all initiator molecules lead to polymer chains which grow at the same average rate at low catalyst concentration, propagation by enolates generated by stoichiometric nucleophilic abstraction of R_3Si groups is unsatisfactory. NMR data[81] confirm that during reactions, prior to benzylation with benzyl bromide,[44] formation of hydroxy-ended polymer by reaction of the living oligomer with benzaldehyde[44] (equation 2.28) and formation of terminal amide functions after reaction with phenyl isocyanate,[81] the silyl ketene acetal function is retained.[44]

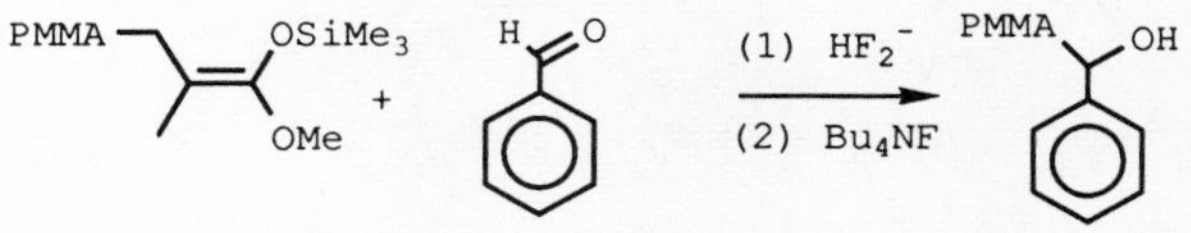

$$\text{PMMA} \overset{OSiMe_3}{\underset{OMe}{\diagdown}} + \overset{H}{\underset{}{\diagdown}} \overset{O}{\underset{}{\diagup}} \xrightarrow[\text{(2) } Bu_4NF]{\text{(1) } HF_2^-} \text{PMMA} \overset{OH}{\underset{}{\diagdown}} \quad (2.28)$$

Two possible dissociative mechanisms for nucleophilic catalysis have been advanced:[81,82] reversible dissociation (Scheme 2.7) and irreversible dissociation and subsequent exchange of trialkysilyl groups (Scheme 2.8).

Reversible dissociation
Evidence against reversible dissociation is that, on polymerization of MMA

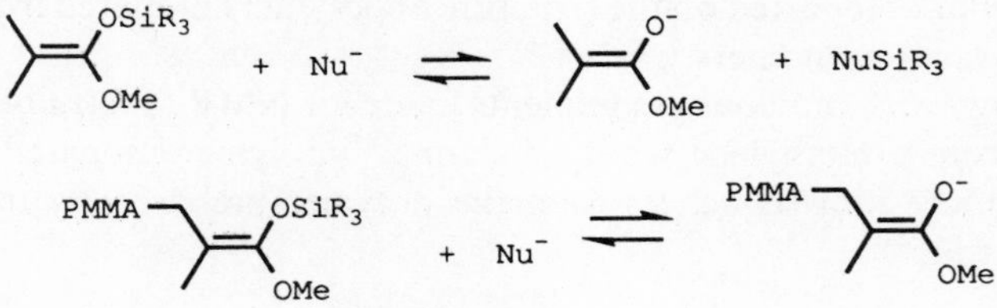

Scheme 2.7

initiated by phenyldimethylsilyl initiator (**41**) in THF (with a trace of acetonitrile) in the presence of an equimolar concentration of tolyldimethylsilyl fluoride, with $TASHF_2$ as catalyst, less than 5% of the tolyldimethylsilyl groups were incorporated into the oligomer.[82] The reaction was stopped at the desired time by adding silver nitrate to sequester the HF_2^-; $TASNO_3^-$ does not catalyse GTP under these conditions.[83] Similarly, using $TASMe_3SiF_2$ as catalyst and adding the tetracoordinated spirosilane (**42**) (which irreversibly coordinates fluoride to form an inert pentacoordinate silicon[17,81,84]) to quench the reaction, it was necessary to go to low temperatures to prevent exchange; about 10% of the tolyldimethylsilyl groups were incorporated into the PMMA in 5 min. at -90 to $-95\,°C$.

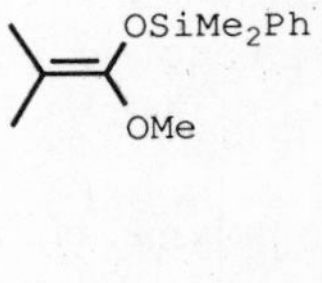

(**41**) 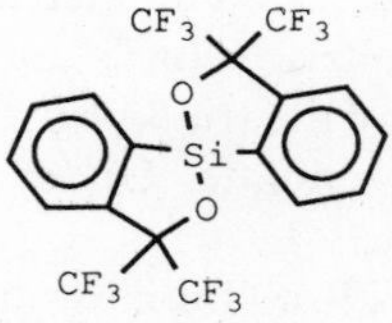   (**42**)

These results suggest that, under the usual polymerization conditions, the back reaction in the first equilibrium in Scheme 2.7 does not occur.

Irreversible dissociation

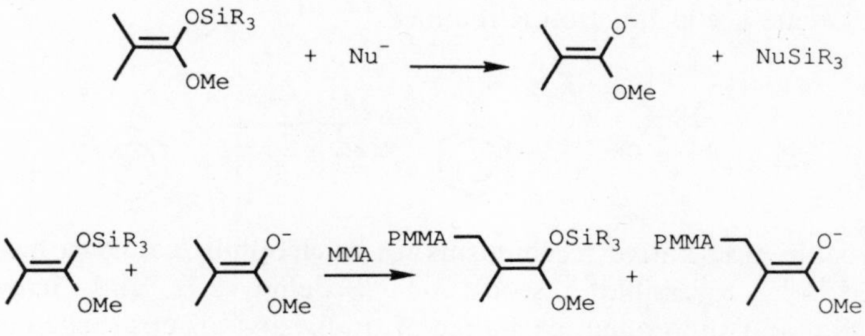

(**43**)   (**44**)

Scheme 2.8

In irreversible dissociation, SiMe$_3$ residues transfer rapidly between species (**43**) and (**44**); (**44**), the active species in anionic polymerization and susceptible to chain termination, is present in only trace quantities (equal to the initial catalyst concentration).

To test for irreversible dissociation, an equimolar mixture of living oligomers of PMMA with tolyldimethylsilyl end-groups ($\bar{M}_n = 2580$) and PBMA with phenyldimethylsilyl end-groups ($\bar{M}_n = 1890$) in THF was reacted with BMA at $-90\,°C$ in the presence of TAS trimethyldifluorosiliconate as catalyst; the reaction was quenched with spirosilane (**42**). PMMA was selectively precipitated into petroleum ether at $25\,°C$ under stringent conditions to retain the silyl ketene acetals. The PMMA extracted, with an additional four units of BMA per chain (approximately), had only tolyldimethylsilyl residues; 15% phenyldimethylsilyl groups in the polymer would have been detected. Thus, on the time scale of the second BMA polymerization, no significant interchange of trialkylsilyl groups between chains occurred.[82] Similar results were obtained with HF_2^- catalyst at room temperature.

The above experiments do not rigorously exclude dissociation for carboxylate catalysts nor the possibility that silyl fluoride cannot stray far before returning to the chain end. However, the current position is that propagation (and initiation) proceeds via intramolecular silyl transfer in which the silyl group is transferred from the ultimate unit of the propagating chain (or initiator) to the incoming monomer. The reaction is assumed to involve hypervalent silicon species. Thus, it is presumed, the nucleophilic catalyst coordinates with the silyl ketene acetal to form a pentacoordinate silicon species that weakens the Si–O bond. An incoming monomer then adds in a non-concerted fashion, the C–C bond forming first, then the silicon group transferring. Between monomer additions the nucleophile must dissociate from the silicon and associate with another species.

In view of the high rates of polymerization which can be achieved, remembering that the concentration of actual propagating chains is no greater than the catalyst concentration, the transfer of the activating nucleophile from chain to chain must occur extremely readily. Thus a mechanism for chain growth for GTP may be written schematically as in Scheme 2.9.

$$R_n\text{-Si} + Nu \underset{}{\overset{K}{\rightleftharpoons}} R_n\text{-Si-Nu}$$

$$R_n\text{-Si-Nu} + M \xrightarrow{k_p} R_{n+1}\text{-Si-Nu}$$

Scheme 2.9

This mechanism presumes that a pentacoordinate silicon species (R_n–Si–Nu) is the active species in propagation. Supporting evidence is that the stable pentacoordinate silicate (**45**) (formed by reacting spirosilane (**42**) with the lithium enolate of methyl isobutyrate,[81,82] initiates polymerization of MMA without intervention of additional catalyst.

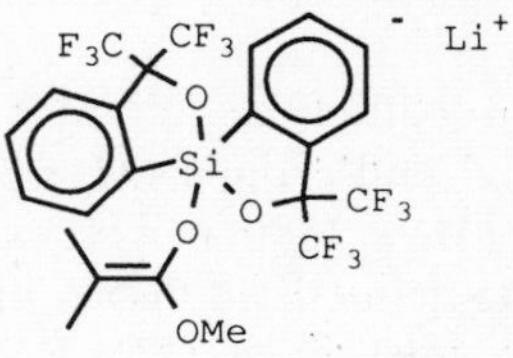

(45) 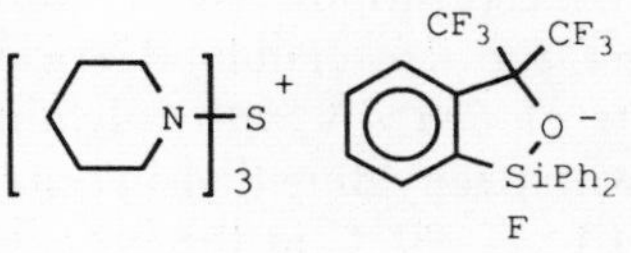   (46)

Using conformational energy calculations, Wei and Wnek concluded that, after coordination of the nucleophile to silicon, the first step is the formation of the new C–C bond.[76] Unpublished *ab initio* calculations led to essentially the same conclusions.[85] From early studies of molecular weights of mixtures of living oligomers of different molecular weights, stirred overnight in the presence of TASHF$_2$, it was concluded that depropagation of active chains does not occur in GTP;[86] there is now evidence to the contrary.[87]

2.3.1.2 *Initiation and propagation* Brittain studied the kinetics of initiation and propagation using stopped-flow techniques.[87] By synthesizing the initiator (10) ($n = 0$) and intermediates (47) ($n = 1$) and (48) ($n = 2$) (n is the number of monomers added to initiator) as pure compounds it was possible to determine the rates of the processes in Scheme 2.10.

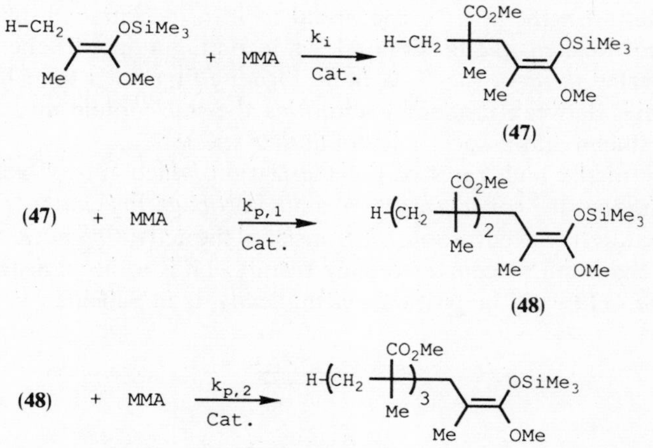

Scheme 2.10

By studying initial rates of reactions of initiator, for example, it could be assumed that $[0] \approx [0]_0$ (species are identified by value of n). For reaction Scheme 2.11, assuming rapid equilibration of catalyst and initiator, Brittain determined that monomer (M) concentration and catalyst (Cat) concentration

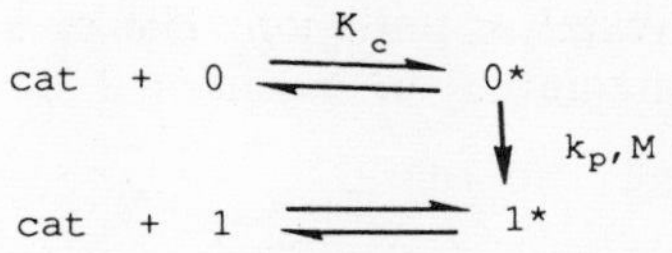

GROUP TRANSFER POLYMERIZATIONS

$$\text{cat} + 0 \underset{}{\overset{K_c}{\rightleftharpoons}} 0^* \downarrow k_p, M$$
$$\text{cat} + 1 \rightleftharpoons 1^*$$

Scheme 2.11

are given by equation 2.29.

$$\ln([M]_0/[M]) = \ln([0]_0/[0]) = k_i K_c [M][0][\text{Cat}] t \qquad (2.29)$$

Schemes equivalent to 2.11 were assumed for species (**47**) and (**48**) and equations corresponding to equation 2.29 were derived with values of $k_{p,1}$, $k_{p,2}$ and [1] and [2] replacing k_i and [0] appropriately. On this basis reactions are first order with apparent rate coefficients k_{app} where, for equation 2.29, $k_{app} = k_i K_c [0][\text{Cat}]$.

Although, as written, the reaction scheme predicts first-order kinetics in [Cat] there is no *a priori* reason for assuming this mechanism. Brittain determined values of k_{app} and, for equation 2.30, determined values of n conventionally. The results for addition of first monomer to initiator are given in Table 2.1.

$$k_{app} = k_i K_c [0][\text{Cat}]^n \qquad (2.30)$$

Thus, Scheme 2.11 is not adequate in all cases. The second-order kinetics in bifluoride, it was suggested, might arise because the transition state involves either (a) a 2:1 $HF_2^-/(\mathbf{10})$ hexavalent silicon complex or (b) a 1:1 fluoride/(**10**) complex, as in equilibrium equation 2.31, in which the bifluoride effectively acts as a source of fluoride. Distinction between (a) and (b) was not achieved but using the tris(piperidino)sulphonium fluorosiliconate (**44**), it was shown

Table 2.1 Kinetic data for polymerizations of MMA using various catalysts

Anion	Cation	n	Concentration[†]	$k_i K_c^{‡} = k_{app}/[\text{cat}]^n[0]_0$
Bifluoride	tris(piperidino) sulphonium	2	$2 - 20 \times 10^{-5}$	9×10^6
Bifluoride	TAS (10% CH_3CN)	2.1		
Fluoride	tris(piperidino) sulphonium (from Fluorosiliconate)	1	$1 - 10 \times 10^{-3}$	
Benzoate	tris(piperidino) sulphonium	1		1.8×10^3
Bibenzoate	tetrabutylammonium	0.3	$1 - 12 \times 10^{-3}$	168 (34)
Bibenzoate	tris(piperidino) sulphonium	0.3		142

[†] range of catalyst concentrations used.
[‡] $[M]_0 = 0.12$ M, $[0]_0 = 0.25$ M, THF, 25 °C

that fluoride can itself catalyse initiation. The possible existence of higher aggregates of HF_2^-, required in equation 2.31, have been discussed elsewhere.[88]

$$(\mathbf{10}) \; + \; 2TPS^+HF_2^- \; \rightleftharpoons \; \underset{OMe}{\overset{F^- \; TPS^+}{\underset{|}{\text{C}}}}\!\!=\!\!\!\underset{}{\overset{OSiMe_3}{}} \; + \; H_2F_3^- \; TPS^+ \quad (2.31)$$

The benzoate catalyst conforms to first-order kinetics. Bibenzoate obeys fractional order kinetics, possibly because it dissociates to benzoate which is the effective catalyst. Bibenzoate dissociates by 33% in methanol,[89] data for reactions in THF are unavailable.

Different catalysts produce different reaction rates. A direct comparison of, say, values of k_i is not possible because k_i and K_c have not been determined individually. In general, overall rates are $HF_2^- \gg$ benzoate > bibenzoate.

Brittain obtained information on the steps corresponding to addition of second and third monomers. Results for the values of $k_p K_c$ for (**47**) and (**48**) are presented in Table 2.2. The values of K_c for various species (using the same catalyst) are probably comparable, and hence the data are consistent with the production of polymers with narrow molecular weight distributions ($k_i \geqslant k_p$) when termination is negligible.

It should be noted that, in all the above studies catalyst concentration was low. The nature of the cation has little influence on rates of reaction.

2.3.1.3 *Termination* Termination was studied in two ways: (a) in mixtures of oligomeric silyl ketene acetals, as produced in polymerization of MMA and (b) of a specific oligomeric silyl ketene acetal.[90] For case (a) the overall composition of the mixtures was 3:1 MMA:initiator.

Brittain and Dicker followed changes in concentrations of specific species. The silyl ketene acetal (**49**) absorbs at $v_{C=C} = 1687 \, cm^{-1}$ and the carbonyl of the substituted cyclohexanone (**50**) at $v_{C=O} = 1740 \, cm^{-1}$; the latter arises from the total concentration of cyclic ketones (from $n \geqslant 3$). Data obtained with independently synthesized pure (**49**) with $n = 3$, showed the variation in concentration of the cyclic trimer from the living trimer. (Note: this termination process does not proceed in the absence of catalyst and the active

Table 2.2 Values of kinetic parameters for MMA polymerizations

Catalyst	TBABBz	TPSBz
$k_i K_c$	34 ± 1	1800 ± 150
$k_p, 1K_c$	48 ± 1	1600 ± 400
$k_p, 2K_c$	28 ± 1	†

TBABBz = tetrabutylammonium bibenzoate; TPSBz = tris(piperidino)sulphonium benzoate.
†not determined.

Table 2.3 Data on kinetics of termination by MMA polymerizations

Ratio [MMA]:[50][a]	Rate of formation of cyclic ketone $l\,mol^{-1}\,s^{-1}$	Total rate of loss of living ends $l\,mol^{-1}\,s^{-1}$
2:1	0.008	0.011
3:1	0.009	0.010

[a] **(49)** with $n = 0$

species involved in termination is probably the species which participates in propagation).

$$H\text{-}(CH_2\text{-}C(CO_2Me)(Me))_n\text{-}C(Me)(OMe)=C(OSiMe_3) \xrightarrow[(-OMe)]{Cat, k_{t,n}} H\text{-}(CH_2\text{-}C(CO_2Me)(Me))_{n-2}\text{-}[\text{cyclic ketone with } CO_2Me, Me, CO_2Me, Me] \quad (2.32)$$

(49)   (50)

Mixtures of MMA with MTS and tetrabutylammonium bibenzoate catalyst (MMA:MTS 2:1, equivalent to $n = 3$ on average), reacted overnight and gave the cyclic trimer in 85% yield. Rates of formation of cyclic ketone and loss of silyl ketene acetal, for specific systems, are summarized in Table 2.3. Cyclic ketones are the major products (73–90%) arising from destruction of the silyl ketene acetal and it was concluded that equation 2.32 is the major inherent termination reaction in GTP. This reaction, a catalysed nucleophilic attack of silyl ketene acetal on a backbone ester, is analogous to the major termination reaction in anionic polymerization of methacrylates (cf. Scheme 2.1). Whether the shortfall in formation of cyclic structures from the silyl ketene acetal is indicative of another termination reaction or of experimental error is uncertain. (Section 2.2.6.1).

In a mixture of MMA and MTS (3:1) PMMA oligomers with degree of polymerization up to 12 formed within minutes and, after quenching, could be identified by gel permeation chromatography. Even at that stage cyclic trimer could be identified specifically and there was a dip at $n = 3$ in the distribution of species. For a reaction allowed to proceed for one day before quenching, all species were transformed to the cyclic product. The distribution of species was also totally changed and the cyclic trimer dominated the distribution. Thus depropagation can occur in the 'living' system and species which revert to the trimer probably terminate as such.

By comparing rates of cyclization for the trimer with rates of propagation it was concluded that $k_{t,3}$ is an order of magnitude greater than $k_{p,3}$. It was also found that for oligomeric systems the termination rate using the bifluoride (TPSHF$_2$) was 7000 times faster than if bibenzoate (tetra-butylammonium) was used. Thus both propagation and termination are slower with bibenzoate.

Brittain and Dicker also compared rates of group transfer and anionic polymerizations of MMA in THF at 25 °C.[90] They found that the ratio of $k_{p,2}/k_{t,3}$ for GTP is thirty times greater than for anionic polymerization; they concluded that GTP of methacrylates is 'more living' than anionic polymerization.

2.3.1.4 *Chain transfer* The other major common reaction in polymerization systems is chain transfer. Hertler established that carbon acids with $pK_a < 25$ are transfer agents in GTP;[91,92] carbon acids with $pK_a < 18$, e.g. malonic esters, terminate GTP. Known transfer agents (S–H) donate a hydrogen to the propagating chain and the residue is silylated, equation 2.33; species S–SiMe$_3$ is an efficient initiator. Reaction 2.33 must be catalysed.

$$\text{MeO-C(Me)=C(OSiMe}_3\text{)} + \text{SH} \longrightarrow \text{~~C(Me)(H)(CO}_2\text{Me)} + \text{S-SiMe}_3 \quad (2.33)$$

The main uses of chain transfer are to reduce the molecular weight of polymer to a desired level and to introduce required terminal units. An additional use in GTP is to reduce the amount of silyl ketene acetal required to produce polymer of low or moderate molecular weight. The molecular weight reduction is described by equation 2.34,

$$\frac{1}{\bar{P}_n} = \frac{1}{\bar{P}_{n,\infty}} + C_s \frac{[\text{S–H}]}{[\text{M}]} \quad (2.34)$$

where $\bar{P}_n$ is the degree of polymerization and $\bar{P}_{n,\infty}$ is the degree of polymerization in the absence of transfer agent. An equation of the form of equation 2.34 is widely used in free-radical polymerization; C_s is defined as the ratio of the rate coefficients for reaction of the propagating radical with transfer agent and with monomer and has values from 10^5 to 10^{-5}. In free-radical polymerization chains are initiated and dead polymer is formed continuously and $\bar{P}_n$ and $\bar{P}_{n,\infty}$ are instantaneous values. In GTP, where conversions approaching 100% are achieved, C_s is defined and used differently. $\bar{P}_{n,\infty}$ is the final degree of polymerization in the absence of transfer for the initiator concentration and monomer conversion achieved ($\bar{P}_{n,\infty} = \{[\text{M}]_0/[\text{I}]_0 + 1\}$). C_s is the fraction of transfer agent present which is involved in reaction and which reinitiates polymerization; C_s has a maximum value of unity.

Effective transfer agents are the arylacetonitriles, benzyl cyanide ($C_s = 0.9$), α-phenylpropionitrile and 2-(β-naphthyl)propionitrile (the first and last named require high levels of catalyst to maintain high rates of polymerization; benzyl cyanide gives less than quantitative conversion of monomer), and the 2-aryl acetates, methyl phenylacetate, methyl (*p*-methoxyphenyl)acetate and methyl α-phenylpropionate. These esters are less efficient transfer agents than the arylacetonitriles. The additional alkyl substituent possibly reduces the rate of the anion-catalysed rearrangement of the intermediate silyl ketene acetal to a less reactive α-silyl ester.[40] Other chain transfer agents are indene, fluorene, α-phenylpropiolactone and γ-thiobutyrolactone. Polymerizations with indene are slow and monomer conversions are less than quantitative; calculated values of C_s vary with the catalyst used.

Mechanisms of transfer by aryl-substituted nitriles are unknown but it has been suggested that N-silyl ketene imines (51) may be the active species formed; transfer possibly leads to (51) as a kinetically controlled product.[91,93] Silylation of phenyl propionitrile gave the C-silylated (52) which even in the presence of high catalyst concentration gives only low rates of polymerization and is unlikely to be the active intermediate; (51) and (52) might interchange in the presence of catalyst.

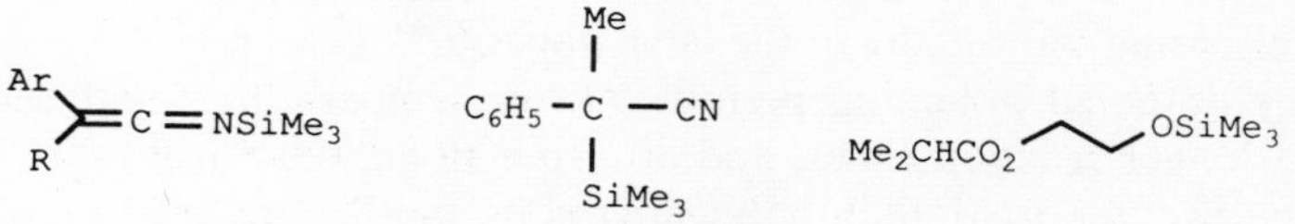

(51)   (52)   (53)

Using the protected hydroxy initiator (14) the proportions of chains derived from initiator and from transfer agent residues were identified. These proportions may vary with the catalyst used. In the case of α-phenylpropionitrile, transfer does not involve reaction of the active intermediate from transfer with the hydrogen in isobutyrate residues in the dead polymer.[91] The chain transfer agent (53) itself provides a protected hydroxyl label and its use gives terminal functionality by chain transfer. Comparable experiments using indene established that the proportions of chains derived from initiator and transfer agent varied with the catalyst used; the reaction is presumed to be as in equation 2.35 and, in support, it was shown that (54) will initiate polymerization.

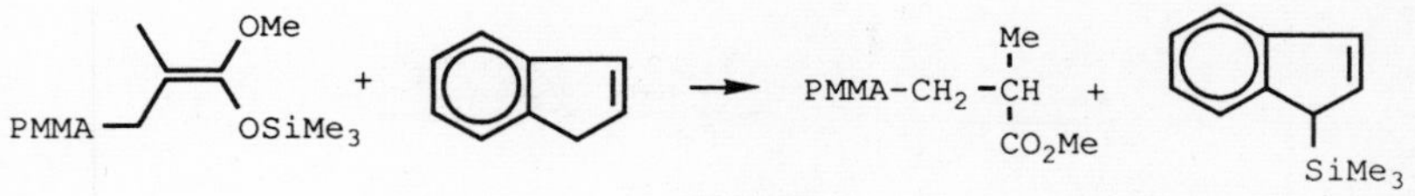

(54)   (2.35)

2.3.1.5 Batch polymerizations The above studies provide an insight into the major mechanistic and kinetic features of GTP. These features, however, have not been combined into a composite mechanism and tested against experimental data and there are indications that the complete mechanism is, in fact, more complex.

Müller and coworkers studied batch polymerizations in which solutions of catalyst (TASHF$_2$) in THF and initiator (MTS) in MMA were mixed rapidly.[75,94-98] These workers adopted a simple kinetic scheme equivalent to Scheme 2.11 without termination, and derived the integrated rate equation 2.36 in terms of [monomer],[94] [P*] is given by equation 2.37. K^* is the equilibrium constant for formation of reactive complex between catalyst and silyl ketene acetal.

$$\ln \frac{[M]_o}{[M]} = K_p[P^*]t \qquad (2.36)$$

$$[P^*] = \frac{K^*[I]_o[C]_o}{1 + K^*[I]_o} \qquad (2.37)$$

They found the simple mechanism to be deficient, with evidence for termination. For [cat] $\geqslant 3 \times 10^{-5}$ mol l^{-1} ([I]$_o = 10^{-3}$ mol l^{-1}) an order in [monomer] of unity in the early stages of reaction was discovered, but with rates decreasing too rapidly in the later stages.[95,96] For [cat] $< 3 \times 10^{-5}$ mol l^{-1} they observed induction periods (at low temperatures) and decreasing rates at longer reaction times, and an order in [initiator] of -0.27. These workers also reported that polymerizations in the presence of oxyanion catalysts are less prone to termination but found induction periods to be more prominent[98]; Brittain and Dicker did not report induction periods.[90] With such catalysts orders in [catalyst] varied from -0.3 to $+1.0$, reflecting different equilibrium constants for active complex formation.

Rather than invoke a chain termination reaction, Müller and coworkers invoked a deactivation step (Scheme 2.12) in which activated species interact with initiator or unactivated propagating chains to form a species, X, which is capable of sequestering catalyst.[94] In favour of this approach they note that, after decreases in rate at moderate conversions, reactions can be accelerated by addition of catalyst.[98]

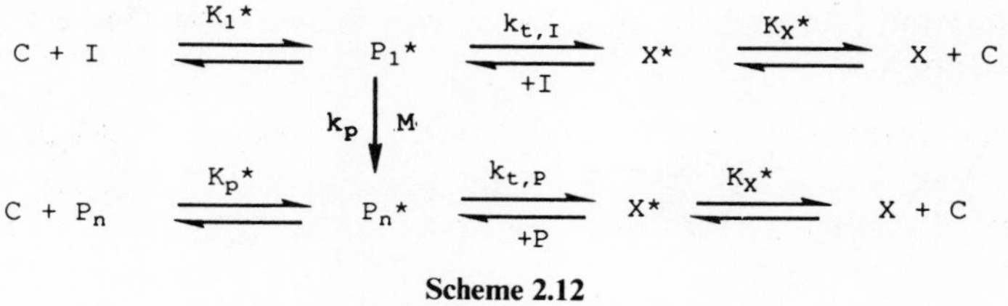

$$C + I \underset{}{\overset{K_1^*}{\rightleftharpoons}} P_1^* \underset{+I}{\overset{k_{t,I}}{\rightleftharpoons}} X^* \overset{K_X^*}{\rightleftharpoons} X + C$$

$$k_p \downarrow M$$

$$C + P_n \underset{}{\overset{K_p^*}{\rightleftharpoons}} P_n^* \underset{+P}{\overset{k_{t,P}}{\rightleftharpoons}} X^* \overset{K_X^*}{\rightleftharpoons} X + C$$

Scheme 2.12

Müller and coworkers derived values of Arrhenius parameters and k_p comparable with those for anionic polymerization of MMA with the larger cations.[95,96] They suggest a mechanism similar to that for anionic polymerization rather than coordination of monomer to a hypervalent silicon which, they suggest, would involve a higher activation entropy and lower frequency factor. Rather than a concerted reaction, equation 2.38, they favour a two-step association process involving addition of monomer to an activated chain-end, in a rate-determining step, and subsequent transfer of the silyl group to the enolate ion so formed, equation (2.39).

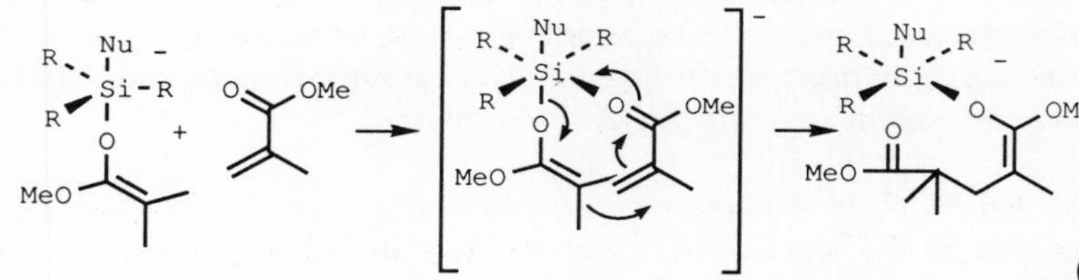

$$(2.38)$$

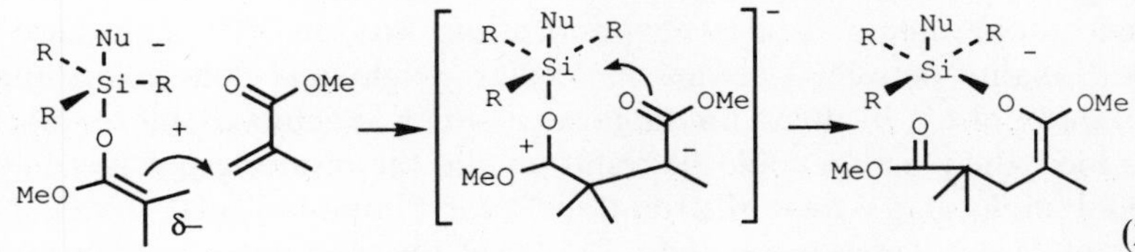

$$(2.39)$$

They do not exclude, but do not favour, a mechanism in which the true propagating species is an enolate ion formed by abstraction of the silyl group by the nucleophile (cf. Reference 99); rate coefficients for propagation are much less than those for free anion propagation. Tacticity data for PMMA prepared by GTP were taken to support a two-step addition process rather than a concerted mechanism. Comparable experiments with t-BMA gave similar results; induction periods were less noticeable except at low temperatures.[97] Termination, however, was found to be more prevalent in GTP than in anionic polymerization. Arrhenius parameters were found to be comparable with those for anionic polymerization and k_p was found to be somewhat higher.

Müller et al. also investigated the molecular weight distributions of polymers formed by GTP in which catalyst was added to a solution of initiator and monomer.[96,98] Under these conditions $[M]_0$ is high compared with monomer-fed reactions. They found that batch processes led to greater polydispersities ($D \approx 1.3$) than monomer-fed techniques ($D \approx 1.03$). Most interestingly, they note that whereas $\bar{M}_{n,\exp} \approx \bar{M}_{n,\text{calc}}$ at the end of a batch polymerization, $\bar{M}_n$ does not increase linearly with conversion. $\bar{P}_n$ and $\bar{P}_w$ show initial rapid increases and, as the polymerization continues, the polydispersity decreases from $D = 4$ to about 1.5 owing to the effect of the relative rates of chain growth and exchange of catalyst, i.e. interchange between activated and unactivated species. In support of this idea they show that D decreases with increasing $\bar{P}_n$ and is always lower for polymerizations in which the stationary value of [M] is lower.[98]

Sitz and Bandermann also reported on batch polymerizations of MMA in THF catalysed by tris(piperidino)sulphonium bifluoride and by tetrabutylammonium cyanide.[100] Both systems gave sigmoidal conversion-time curves (induction periods). They found that in the former case MTS is rapidly isomerized and nearly 25% exists in the C-silylated form; it also converts to the corresponding ester, releasing Me_3SiF and other products. These side reactions decrease the concentration of active propagating species. The reactions are more prevalent at high catalyst concentration but are relatively slow and less significant with Bu_4NCN catalysis.

2.3.1.6 *Reactions in acetonitrile* Polymerizations in acetonitrile (a convenient solvent for less-soluble catalysts such as $TASF_2SiMe_3$), or THF-acetonitrile mixtures, are complex.[39,51,59,60,100] Amongst the complications observed were: incomplete conversions of monomer at 'low' initiator concentration, unless catalyst concentration was 'high';[51] and lack of proportionality between polymer molecular weight and conversion under some conditions;[60] if, after a limiting conversion was achieved, more catalyst was added, the reaction could be restarted and the resulting polymer had a bimodal molecular-weight distribution;[60] for $[\text{initiator}] \geqslant 10^{-3}$ mol l^{-1}, values of $\bar{M}_{n,exp}$ were greater than $\bar{M}_{n,calc}$; at lower initiator concentration molecular weights and limiting conversions passed through a maximum as catalyst concentration increased; values of $\bar{M}_{n,exp}$ decreased with increasing catalyst concentration at [catalyst]/[initiator] > unity;[60] induction periods decreased as [catalyst]/[initiator] increased[39]. Similar results were obtained using Et_4NCN as catalyst[59] but in this case incomplete monomer conversions were observed only at lower initiator concentration than when $TASF_2SiMe_3$ was used as catalyst. These observations were taken to indicate chain termination involving loss of catalyst.

It was suggested that in acetonitrile MMA reacts with the catalyst to produce a series of specific products[39,60] and decreases the amount of catalyst available for polymerization. They also found that acetonitrile undergoes a $CN^{\ominus}$ catalysed reaction with MTS initiator to silylate the solvent and produce Me_3SiCH_2CN.[39,60] It was presumed that this reaction, with the silyl ketene acetal of the propagating chains, is a major cause of termination in these systems. The same group also found that, in the presence of Et_4NCN, MTS isomerizes to its C-silylated form.[60] Similarly, the product of addition of the first monomer to Me_3SiCN initiator isomerizes to 25% of the keto form,[59] equation (2.40).

$$\underset{Me}{\overset{NC-CH_2}{>}}\!\!=\!\!\underset{OMe}{\overset{OSiMe_3}{<}} \quad \rightleftharpoons \quad \underset{COOMe}{\overset{NC-CH_2}{\underset{|}{Me-C-SiMe_3}}} \qquad (2.40)$$

They also found evidence that, in competition with catalysing polymerization reactions, that Et_4NCN catalyst complexes readily with the Me_3SiCN initiator, greatly reducing the amount of catalyst available for catalysis. Thus initiation is slow and, as long as free Me_3SiCN is present, polymerization cannot proceed effectively, giving rise to 'induction' periods in which polymerization proceeds very slowly. The use of mixed initiators demonstrated that Me_3SiCN complexes the catalyst much more strongly than does the silyl ketene acetal of the propagating chain. Although a significant number of features of polymerizations in acetonitrile have been elucidated, the details remain unclear.

2.3.2 *Polymerization of acrylates*

While work on GTP has concentrated on the polymerization of methacrylates, the polymerization of acrylates with fluoride catalysis was hinted at initially; a methacrylate/acrylate diblock copolymer was prepared.[17] However, the polymerization of acrylates has greater attendant problems[40,69] (molecular weight distributions are broader than those of methacrylates); there are also significant differences between anionic polymerizations of acrylates and methacrylates.[101] Lewis acids are the preferred catalysts for acrylate polymerizations.[17,18,40,69] Zinc halides and dialkylaluminium halides (or oxides) have been used and, of the former, the iodide gives better results than the chloride. Mercuric iodide catalysis has since been shown to give improved control over acrylate polymerizations; MTS and (12) are proven initiators.[102]

Anion catalysts (used in 0.1 mol% relative to initiator) are considered to interact with the silicon of the silyl ketene acetal. Lewis acids, used in higher concentrations (aluminium catalysts are used in 10–20 mol% relative to *initiator*), especially the zinc catalysts (used in 10–20 mol% relative to *monomer*),[68] are considered to interact with the monomer and activate it for nucleophilic attack by the silyl ketene acetal.[69] Lewis acids (zinc and aluminium compounds) also modify the kinetics of free-radical polymerization through interaction with acrylate and methacrylate monomers[103–105] and the same compounds have been used to catalyse reactions of silyl enol ethers and silyl ketene acetals.[107–108] In contrast, only low concentrations of mercuric iodide are required (~ 10 mol% based on *initiator*) and it is assumed that this catalyst interacts with the silyl ketene acetal to give a 1:1 complex or that the initiator forms a complex with a dimer of HgI_2, equation 2.41; the latter type of complex (55) has been implicated in E/Z isomerization of silyl ketene acetals catalysed by mercuric iodide.[102,108]

$$\underset{OSiMe_3}{\overset{OMe}{\diagup\!\!\!\!\diagdown}} + (HgI_2)_2 \longrightarrow \underset{\underset{(55)}{OSiMe_3}}{\overset{\overset{\delta-}{I-HgI_2}}{\underset{\delta+}{\overset{I\cdots Hg\cdots OMe}{\diagup\!\!\!\!\diagdown}}}} \qquad (2.41)$$

While aluminium catalysts can be used in acetonitrile mixtures as solvents, the donor solvents used with anion catalysts are to be avoided in Lewis acid catalysed polymerizations; halogenated alkanes and hydrocarbons are preferred.[69] With mercuric iodide catalysis both polar and non-polar solvents give narrow molecular weight distributions.[102] Polymerizations in toluene were slow but gave well-controlled reactions.

Acrylates are more reactive than the corresponding methacrylates and, in a mixture of the two, the acrylate is polymerized virtually exclusively.[69] Side reactions, however, limit the 'livingness' of propagating acrylate chains. After polymerization of ethyl acrylate at ambient temperature using a zinc catalyst, addition of more monomer gave no further polymerization. At ambient temperatures aluminium catalysts give low yields of polymer from MMA but good conversion of acrylates if catalyst is added to a solution of monomer and initiator; monomer feeds give low yields. At $-78\,°C$ the termination process is retarded and living polymers can be obtained. MMA can be polymerized to high conversion at ambient temperature using zinc halide catalysis.[40,69] Mercuric iodide successfully catalysed polymerizations of acrylates and N,N-dimethylacrylamide but not of MMA, acrylonitrile, acrylamide or maleimides.[102] The improved livingness of mercuric iodide catalysed polymerizations allowed near-quantitative end-capping of growing chains by benzaldehyde, equation 2.42. Hydrolysis of the product, when the protected hydroxyl initiator was used, gave α,ω-dihydroxy polymer. Termination with 1,3-dioxalane also gave polymer with Me_3SiO end-group, equation 2.43.

$$\text{PEA}\underset{OSiMe_3}{\overset{H}{\diagup}}\!\!\!=\!\!\!\underset{}{\overset{OEt}{\diagdown}} + PhCOH \longrightarrow \underset{PEA\ \ Ph}{\overset{EtO_2C\ \ OSiMe_3}{\diagup\diagdown}} \quad (2.42)$$

$$\text{PEA}\underset{OSiMe_3}{\overset{H}{\diagup}}\!\!\!=\!\!\!\underset{}{\overset{OEt}{\diagdown}} + \underset{O}{\overset{O}{\square}} \longrightarrow \underset{CO_2Et}{\overset{PEA}{\diagup}}\!\!-\!\!O\!\!-\!\!OSiMe_3 \quad (2.43)$$

Reaction with terephthaldehyde gave limited yields of coupled polymer, in spite of good coupling with the monomeric silyl ketene acetal, indicating the presence of a termination reaction. Accompanying E/Z isomerization in silyl ketene acetals is a slow isomerization from an O-silyl ester (56) to an α-silyl ester (57),[102,109] equation 2.44.

$$\underset{OSiMe_3}{\overset{OMe}{\diagup}} \longrightarrow Me\!-\!\underset{H}{\overset{SiMe_3}{|}}\!-\!CO_2Me \quad (2.44)$$

(56)   (57)

C-silyl compounds are not initiators for GTP catalysed by mercuric iodide and such a rearrangement, which would probably occur more readily in polar solvents, would constitute a termination reaction.[102]

Evidence for the mechanism of a termination reaction in anion-catalysed polymerizations of acrylates was obtained from the reaction between a butyl acrylate oligomer ($\bar{P}_n = 4$), prepared in THF at $0\,°C$ using TASF as catalyst, and *p*-nitrobenzyl bromide at −78 °C. Two products were formed by eliminating Me$_3$SiBr. In one the *p*-nitrobenzyl group was attached to a terminal unit of the chain and in the other at a site along the chain.[40] The suggested mechanisms are represented in Scheme (2.13). The results suggest that isomerization attaches trimethylsilyl groups to internal units of the chain. Because silyl ketene acetals with bulky substituents will not initiate polymerization, isomerization acts as a termination reaction.[40]

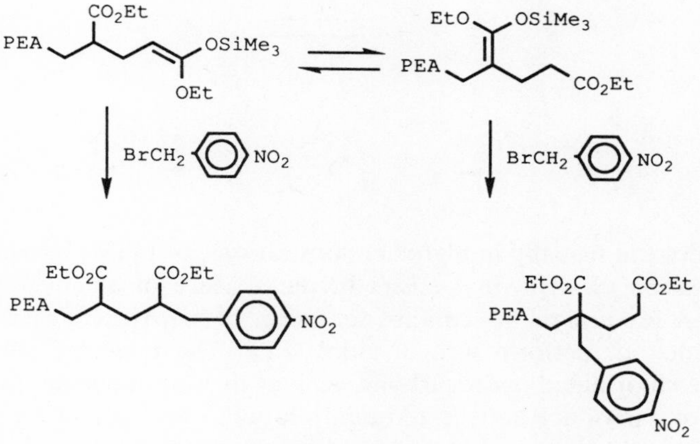

Scheme 2.13

2.4 Aldol group transfer polymerization

Aldol-GTP is an application of a catalysed silyl aldol condensation.[86,110,111] Lewis acid-catalysed reactions of silyl enol ethers with aldehydes and ketones can proceed to high conversion and can give highly selective products.[27–29,112] Addition of an aldehyde to a silyl vinyl ether regenerates an aldehyde, equation 2.45, and provides the potential for a chain growth reaction and a living polymerization. The trialkyl silyl group is transferred from incoming monomer to the active chain end.

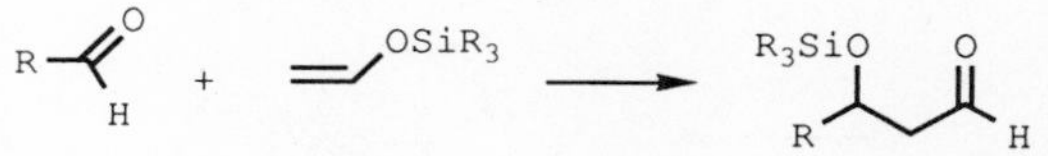

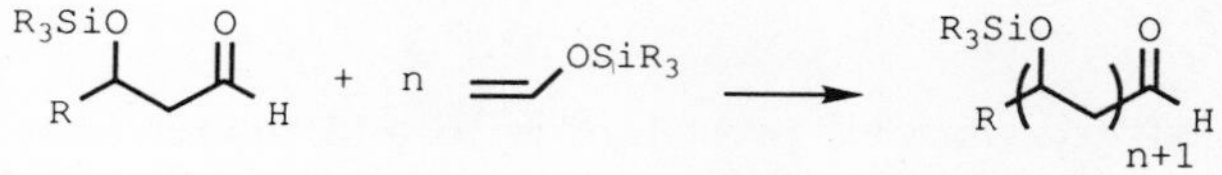

Little work on aldol GTP has been reported.[110,111] However, appropriate choice of reagents gives high conversions of silyl vinyl ethers of polymer, primarily of low molecular weight with narrow molecular weight distributions

($D = 1.02$–1.5, approx.). Zinc halides are the preferred catalysts;[110,111,86] other catalysts give low conversion of monomer. In contrast to GTP, where Lewis acid catalysts are used in high concentrations, low concentrations of zinc halide ($\sim 10\%$ [initiator]) are effective and it has been suggested[110] that the mechanism might involve coordination of halide to initiator, equation 2.47, and subsequent addition of monomer, possibly involving an intermediate β-silyl stabilized carbocation, equation 2.48.

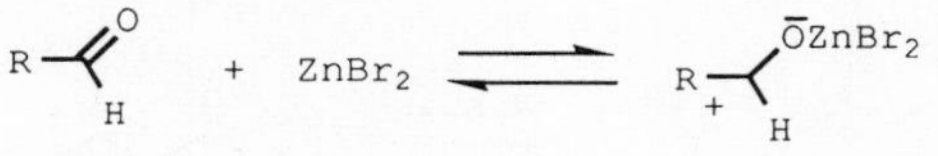

$$R-CHO + ZnBr_2 \rightleftharpoons R-CH(\bar{O}ZnBr_2)^+ \quad (2.47)$$

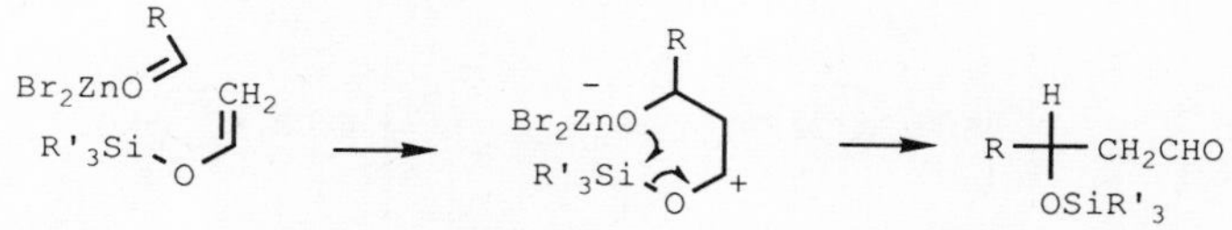

$$(2.48)$$

Zinc halides can (usually in higher concentrations, ~ 1 mM) initiate *cationic* polymerization of silyl vinyl ethers in the absence of aldehyde; yields of polymer are low and the systems are not living.[113–115] Anionic catalysts, used in GTP, do not perform well in aldol GTP. The preferred solvents are unreactive chlorinated hydrocarbons, such as dichloromethane, rather than tetrahydrofuran or acetonitrile; tetrahydrofuran or acetonitrile are necessary if anionic catalysts are employed.[86] Better control of product molecular weight was achieved using, *t*-butyldimethylsilyl vinyl ether as monomer in preference to trimethylsilyl vinyl ether, which correlates with the greater hydrolytic stability of the former.[29,116,117] Diphenylmethylsilyl, triethylsilyl and isopropyldimethylsilyl vinyl ethers have also been employed.[86] Aromatic aldehydes are the preferred initiators. Benzyl halides and acetals can initiate polymerization and 1,4-bis(bromomethyl)benzene and terephthaldehyde have been suggested as difunctional initiators,[86,111] equations 2.49 and 2.50.

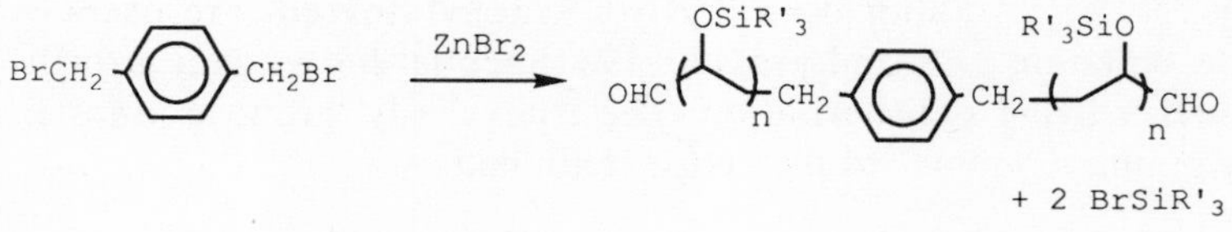

$$(2.49)$$

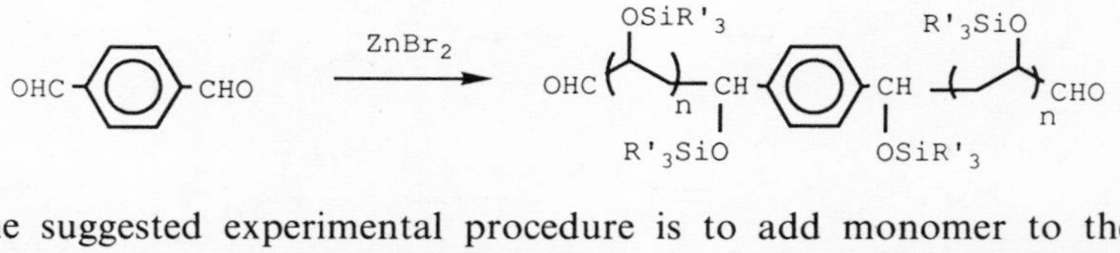

$$+\ 2\ BrSiR'_3$$

$$(2.50)$$

The suggested experimental procedure is to add monomer to the other reagents under argon. Reactions are reported to proceed satisfactorily over the temperature range -80 to $70\,°C$.

While aldol GTP has an associated termination reaction, polymerizations presumably live long enough for the products to be coupled with living polymers with terminal silyl ketene acetal groups (Section 2.5.2). Aldol GTP is a potential alternative to cationic polymerization which does not give living polymers.[113-115] Living cationic polymerizations of vinyl ethers has now been reported[8] but apparently has not been applied to silyl vinyl ethers. A potential interest in aldol GTP is the synthesis of well-defined samples of poly(vinyl alcohol) by removal of the trialkylsilyl groups from the polymer, equation (2.51).

$$\sim\!\!\!-\!\!\!\left(\mathrm{CH_2-\underset{OSiMe_3}{\overset{}{C}H}}\right)_n\!\!\!\sim \quad \xrightarrow[\text{MeOH}]{\text{F}^-/\text{THF}} \quad \sim\!\!\!-\!\!\!\left(\mathrm{CH_2-\underset{OH}{\overset{}{C}H}}\right)_n\!\!\!\sim \qquad (2.51)$$

Whichever silyl vinyl ether monomer is used, poly(vinyl alcohol) can be generated by refluxing the polymer in THF/methanol with tetrabutyl ammonium fluoride or hydrogen chloride.[18,116,117] This contrasts with the free-radical polymerization of vinyl acetate to branched products and hydrolysis to complex distributions of poly(vinyl alcohol) molecules.

The polymerization of 1-butadienyloxytrimethylsilane, with benzyl chloride as initiator, has been reported;[118] the product was soluble, unlike that formed in the presence of benzaldehyde. The product was a living polymer containing *trans* double bonds in the main chain.

2.5 Copolymers

We identify three approaches to the use of GTP in copolymer synthesis.

1. random copolymerization of monomer mixtures;
2. sequential polymerization of two or more monomers;
3. coupling of performed terminally active polymers.

The relative reactivities of different species limit the flexibility of approaches (1) and (2). In synthesizing binary copolymers (from an acrylate and a methacrylate) we are concerned with the relative rates of the four propagation reactions in Scheme 2.14.

It is usual to define reactivity ratios which control instantaneous copolymer compositions from monomer feed compositions. The reactivity ratios for acrylate and methacrylate may be defined as $r_A = k_{p,a}/k_{p,b}$ and $r_M = k_{p,c}/k_{p,d}$, respectively; these ratios determine the sequence distributions of monomer units in the copolymer. During random copolymerization the copolymer and monomer feed compositions are usually different and the copolymer composition drifts with reaction time. In a living polymerization, which GTP approximates, this drift occurs along the length of a polymer chain.

For addition of monomer in initiation a methacrylate active end is generally more reactive than an acrylate active end i.e. $k_{p,c} > k_{p,b}$ and $k_{p,d} > k_{p,a}$. Acrylates are much more reactive in propagation than methacrylates even

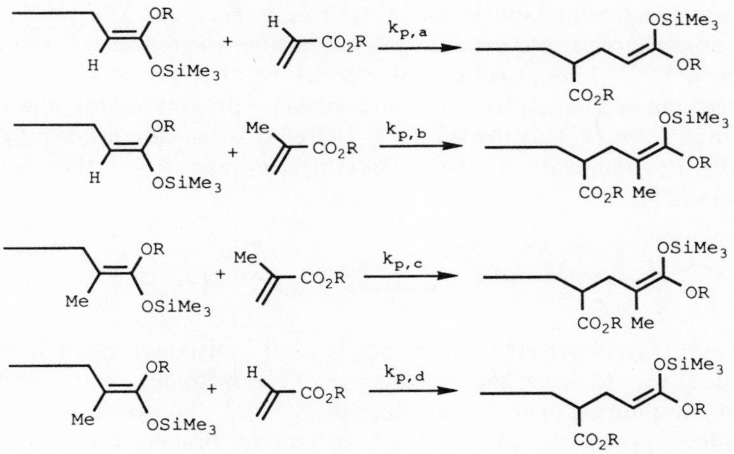

Scheme 2.14

using anion catalysts; $k_{p,a} \gg k_{p,c}$. Also, the termination of acrylates is faster than that of methacrylates. In consequence, a chain with a terminal active acrylate unit will not readily add a methacrylate. Thus, in polymerization of a mixture of monomers the acrylate will be consumed rapidly, leaving the methacrylate largely unreacted; an acrylate homopolymer will essentially be formed. Hence, to form methacrylate–acrylate block copolymers the methacrylate block should be produced first. The optimum catalysts for polymerization of the two types of monomer are not the same so that addition of acrylate to a living polymethacrylate containing an anion catalyst will not necessarily give as narrow a molecular weight distribution as if the acrylate was polymerized under optimum conditions.

2.5.1 Random copolymers

Random copolymers of MMA, BMA, and glycidyl methacrylate (GMA) have been prepared with narrow molecular weight distributions ($D = 1.1$)[40] (the BMA, GMA copolymer had T_gs at -13 and $+47\,°C$, indicating block copolymer formation and microphase separation of the blocks) and of MMA and BMA.[57] To date the only report of reactivity ratios in GTP is for the copolymerization of MMA and n-butyl methacrylate (nBMA).[119] Copolymers were prepared in THF using MTS as initiator and TASHF$_2$ in acetonitrile as catalyst. Monomer mixtures, added slowly, were left to react to 20% conversion. Reactivity ratios were $r_{MMA} = 0.44 \pm 0.03$ and $r_{nBMA} = 0.26 \pm 0.03$. These values differ from those for free-radical polymerization of the same monomer pair at $60\,°C$ ($r_{MMA} = 0.79$[120] or 1.20[121] and $r_{BMA} = 1.27$) and from estimated limiting values for anionic polymerization.

Random copolymerization of MMA with *p*-vinylbenzyl methacrylate was used to prepare polymers with pendant styryl units susceptible to free-radical polymerization.[68,122]

It is claimed that MMA and 5-methacryloxyvaleraldehyde (protected as its acetal) can be randomly copolymerized to copolymer (**58**) ($\bar{M}_n = 5770, D = 1.4$, 47% MMA) which can be deprotected, equation 2.52, to produce (**59**), a copolymer with pendant aldehyde groups.[86] The aldehyde groups can be used to initiate aldol-GTP of silyl vinyl ethers to produce comb-graft copolymer (**60**) ($\bar{M}_n = 7800, D = 1.96$). Removal of the silyl groups gives a comb-graft copolymer of PMMA and PVA (**61**).[86]

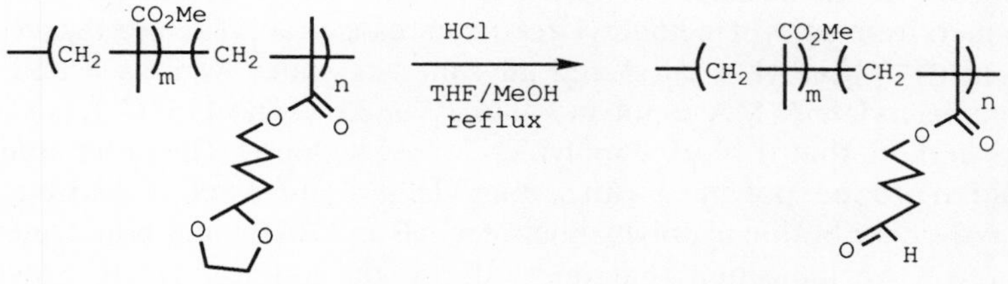

(2.52)

(**58**)　　(**59**)

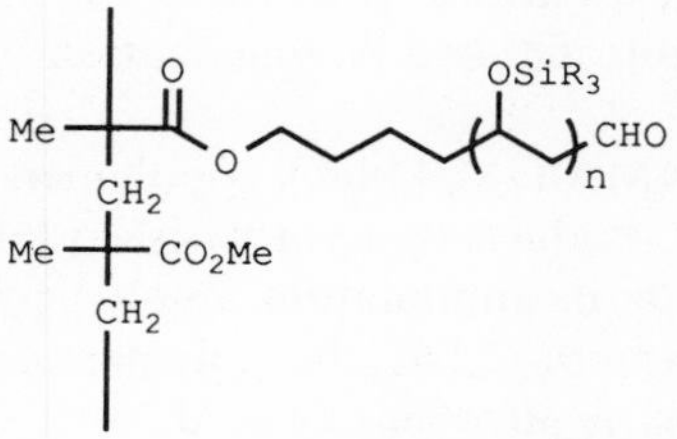

(**60**) 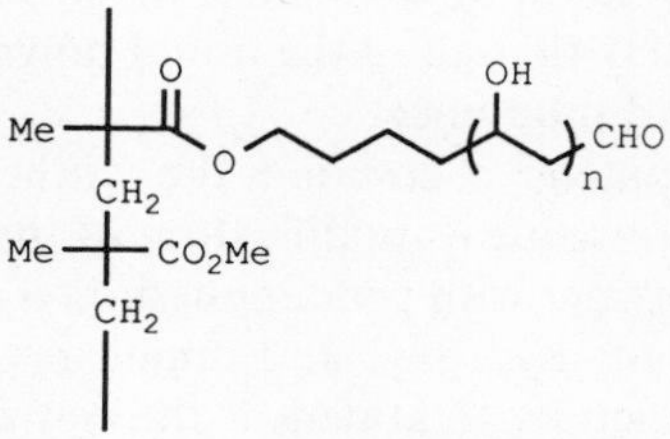　　(**61**)

2.5.2 Block copolymers

The original publication of GTP[17] reports the syntheses of di- and tri-block copolymers based on MMA, nBMA and allyl methacrylate with $D = 1.07$, using TASHF$_2$ as catalyst in THF. The synthesis of a MMA/MA diblock copolymer ($D = 1.9$) using the same system was also reported. An example of initiation of one methacrylate by growing chains of another is the synthesis of diblock copolymers of MMA and lauryl methacrylate.[40,68] AB block copolymers were prepared, using TASHF$_2$ catalysis in THF, polymerizing the monomers in different order to produce copolymers with $D = 1.1$ approximately. Diblock copolymers ($D = 1.05$) of MMA and butyl methacrylate, using TASHF$_2$ (11%) in THF, have been reported.[57]

The difunctional initiator (**62**) was used to prepare BAB block copolymers, polymerizing MMA and hydroxyethyl methacrylate (with the hydroxy group protected by trimethylsilyl) to a random copolymer A and then adding lauryl methacrylate to produce B blocks.[40] Copolymers in which the first block was a random copolymer of MMA and butyl methacrylate and the second was poly(allyl methacrylate)[68] were also prepared.

$$\text{Me}\underset{\text{Me}}{\overset{\text{OSiMe}_3}{\diagup}}\text{O}\diagdown\text{O}\underset{\text{OSiMe}_3}{\overset{\text{Me}}{\diagup}}\text{Me}$$

(62)

Supposed difunctional initiators (**18, 19**) were used to synthesize triblock copolymers from pairs of methacrylates (in one case an acrylate was the second component).[48] Methyl, n-butyl and allyl methacrylates were used and the butyl methacrylate/MMA copolymer had T_gs at 35 °C and 115 °C. It is likely, but not certain, that triblock copolymers were produced. The latter study[48] was said to produce polymers with a unimodal molecular weight distribution. Even so the distribution of polymer between AB and BAB block copolymers is not known. An indication that only 60% of the initiator led to polymer was attributed to the use of a polystyrene calibration for gel permeation chromatography. However, acetonitrile was used as catalyst solvent and at least one hour was left between monomer additions. It is, therefore, highly unlikely that all of the initial polymer chains initiated polymerization of the second monomer.

A patent[123] describes the synthesis of MMA/GMA block copolymers and the subsequent modification of the epoxy residues, by hydrolysis, to form a copolymer with pendant hydroxyl groups or, by ammination, a polymer with pendant hydroxyl and amine residues (Section 2.2.5); the polymer can be modified by treatment with dinitrobenzoic or nitrobenzoic acid.

The other approach to block copolymer synthesis is to prepare and use terminally functionalized blocks. One example is to terminate living PMMA with bromine to incorporate a terminal Br atom[17,18] and to use the functionalized polymer as a macroinitiator.[124] This approach was used to synthesize block copolymers of MMA and styrene, using a redox reaction to remove the terminal Br to produce free radicals in the presence of styrene monomer. This procedure gave block copolymers with a PMMA block with a narrow molecular weight distribution and polystyrene blocks with a broader molecular weight distribution. The copolymer composition distribution indicated a mixture of AB and ABA block copolymers; the AB copolymer might have arisen from chain transfer.

The reverse procedure uses a preformed polymer as an initiator for GTP. This approach has been used[40,68] as depicted in Scheme 2.15. Polycaprolactone ($D = 1.76$) with terminal hydroxy groups was reacted with acryloyl chloride to give a polymer with α,ω-acrylic units. These were converted to silyl ketene acetals by reaction with trimethylsilylcyanide and tetraethylammonium cyanide as catalyst for silylation. Polymerization of MMA was initiated from the terminal residues to give a copolymer (BAB) with $D = 1.78$.

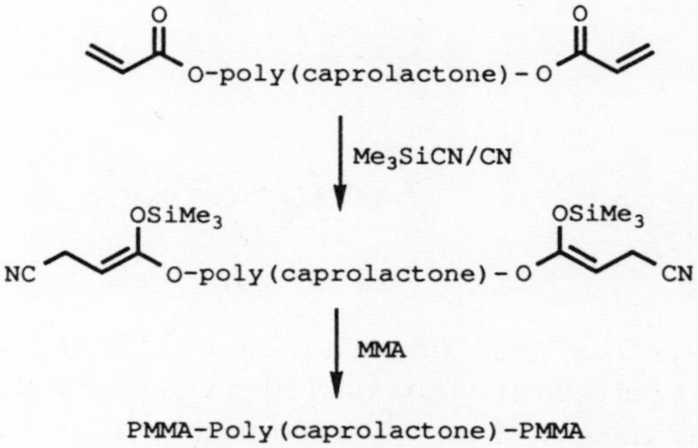

Scheme 2.15

Other polymeric α,ω-diols prepared by GTP have been used to synthesize copolymer by chain extension reactions (see also Section 2.6). An example is chain extension of α,ω-dihydroxy PMMA with bis(p-isocyanatophenyl)methane in THF to form a PMMA-urethane polymer.[125] Polymers with terminal isocyanates will give segmented block copolymers. Polydimethylsiloxane prepared by anionic ring-opening polymerization of hexamethylcyclotrisiloxane and capped with a chlorosilane derivative of allyl methacrylate (Scheme 2.16) was used as a macroinitiator for GTP.[126] The end-group was hydrosilylated (with ethyldimethylsilane and a Wilkinson catalyst) to give a terminal silyl ketene acetal moiety, a macroinitiator for GTP of MMA. The polymer was used to prepare AB block copolymer.[127]

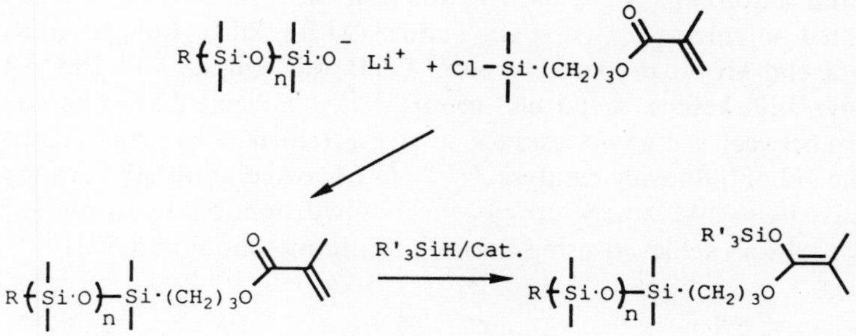

Scheme 2.16

Similarly, block copolymers were synthesized by a combination of metathesis ring-opening and aldol group transfer polymerizations.[128] Aldehydes (and other carbonyl compounds) react with titanium carbenes, equation 2.53, generated from titanocyclobutanes.[129] Thus initiator (**63**) was reacted with norbornene to give a living polynorbornene, the reactive end-group of which was reacted with excess terephthaldehyde to give polymer (**64**).

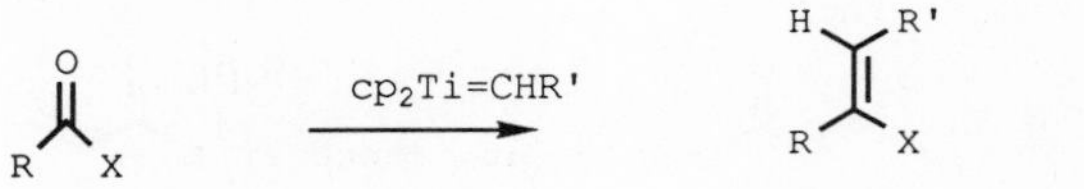

$$\underset{R}{\overset{O}{\underset{\|}{C}}}\;X \quad\xrightarrow{cp_2Ti=CHR'}\quad \underset{R}{\overset{H}{\underset{}{C}}}=\underset{X}{\overset{R'}{\underset{}{C}}} \qquad (2.53)$$

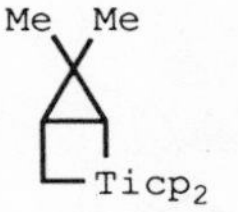

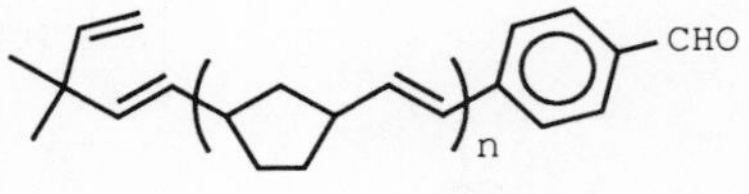

(**63**) Me Me / Ticp₂ cyclopropane

(**64**) polymer chain with CHO end group

Polymer (**64**) was then used to initiate aldol-GTP and so make block copolymers. Tert-butyldimethylsilyl vinyl ether was used as the monomer and zinc chloride as catalyst. The polynorbornene-poly(*tert*-butyldimethysilyl vinyl ether) copolymers had narrow molecular weight distribution ($D = 1.1–1.6$); block copolymers were made using exo-dicyclopentadiene as monomer for metathesis polymerization.[128] Aldehydic end-groups were reduced to the corresponding alcohol by reaction with sodium borohydride to prevent subsequent cleavage of β-hydroxyaldehyde in the known retrograde aldol reaction in the presence of nucleophiles or bases.[130] Then cleavage of the trialkyl silyl group (Section 2.4) gave hydrophobic–hydrophilic block copolymers. The PVA blocks were acetylated to produce soluble block copolymers. The thermal transition behaviours of the polymers were investigated.

2.5.3 Coupling

GTP and aldol-GTP were used to make a block copolymer by coupling preformed polymers. Poly(silyl vinyl ether) (with $OSiMe_2Bu^t$ group) with an aldehyde end-group, prepared by aldol-GTP, was coupled to a PMMA with an active silyl ketene acetal end-group,[110,111] Scheme 2.17. The coupling reaction between these two residues, to give a Reformatsky product, proceeds with the aid of bifluoride catalysis.[27,29,131] Cleavage of the silyl groups from the poly(silyl vinyl ether), to produce a hydrophobic/hydrophilic block copolymer, was achieved using fluoride ion in methanol and THF[116].

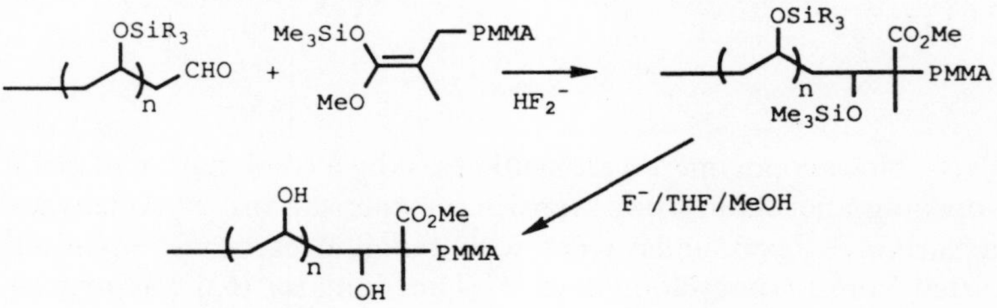

Scheme 2.17

2.5.4 Other routes

Aldol GTP can be used to produce a wide variety of copolymers. For example, if two silyl vinyl ethers, with different propensities to hydrolysis, are used sequentially block copolymers with one hydrophilic and one hydrophobic block can be synthesized by selective cleavage of trialkyl silyl groups. Thus, after converting the aldehyde end-groups in block copolymer (**65**), to ester, by reacting with MTS (equation 2.54), the trimethylsilyl groups can be selectively cleaved with the aid of fluoride to give a hydrophilic/hydrophobic block copolymer (**66**)[29,116,117]

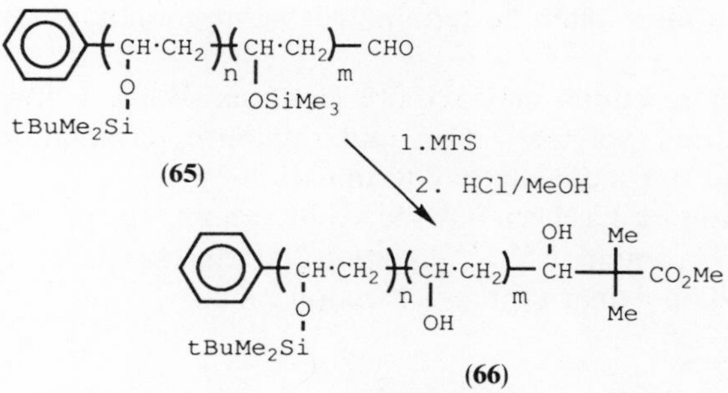

(2.54)

Novel block copolymers can be prepared by terminating 'living' polymers produced by anionic polymerization to produce macroinitiators for aldol-GTP. For example, in Scheme 2.18, the intermediate and final copolymers each have two T_gs.[86]

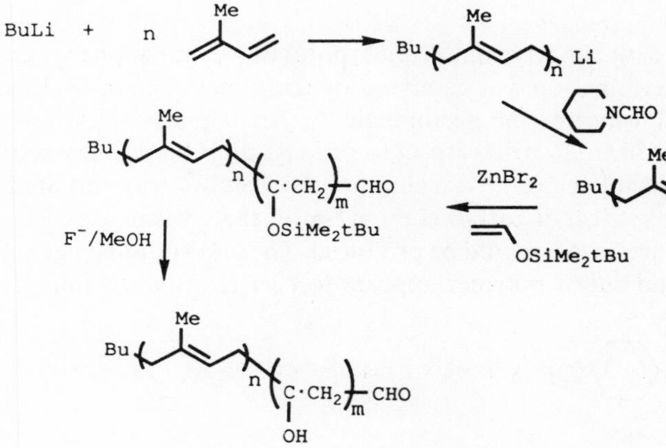

Scheme 2.18

Similarly, it is claimed,[86] by synthesizing polystyrene with a terminal aldehyde group, a block copolymer of polystyrene and poly(*t*-butyldimethylsiloxy-vinylether) ($\bar{M}_n = 9600$, $D = 1.5$) and hence a polystyrene/poly(vinyl acetate) block copolymer can be prepared.

66   NEW METHODS OF POLYMER SYNTHESIS

2.6 Telechelics

Living polymerizations permit the synthesis of telechelics, i.e. polymers with specific terminal units. Reactive terminal units may be introduced by using selected initiators or termination reactions (See Chapter 6). Section 2.2.3 describes the introduction of hydroxyl and carboxyl groups through the use of initiators with such groups trapped by trimethylsilyl residues; see also Section 2.3.1.4. This section considers the use of selected termination reactions; the inherent termination reaction, if allowed to proceed, reduces the proportion of chains functionalized. For optimum functionality a polymerization must often be terminated before monomer consumption is complete.

Termination reactions may require catalysis. While a weak catalyst is chosen to sustain polymerization and minimize termination, a stronger catalyst is used to ensure effective termination.

Terminal bromine has been introduced by reacting the propagating species with molecular bromine,[18,44,124] equation 2.55; the telechelics so formed were used in block copolymer synthesis (Section 2.4.2).

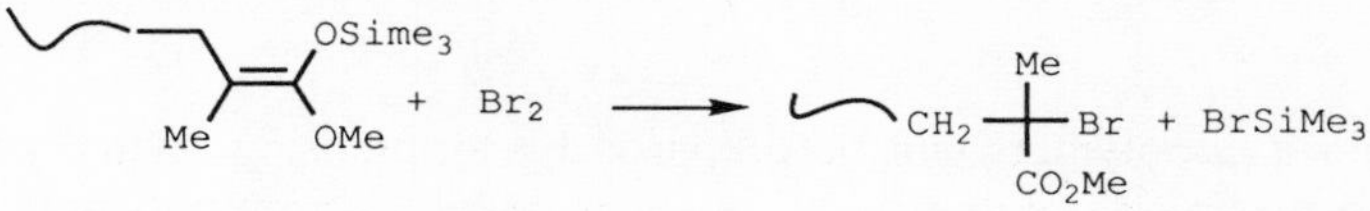

$$\text{~~~C(Me)=C(OSiMe}_3\text{)(OMe)} + Br_2 \longrightarrow \text{~~~CH}_2\text{-C(Me)(CO}_2\text{Me)Br} + BrSiMe_3 \quad (2.55)$$

$$\text{~~~C(Me)=C(OSiMe}_3\text{)(OMe)} + \text{Ph-CH}_2\text{Br} \xrightarrow{\text{Cat.}} \text{~~~CH}_2\text{-C(Me)(CO}_2\text{Me)Bz} \quad (2.56)$$

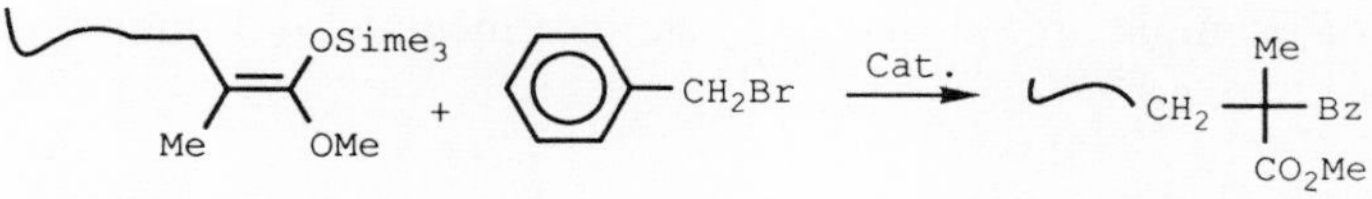

Termination with benzyl bromide incorporates a terminal phenyl group,[44] equation 2.56; termination was catalysed by tris(dimethylamino)sulphonium difluoro-trimethylsilicate in acetonitrile.[18] Analagously, 1,4-bis(bromomethyl)benzene (50 mol% with respect to propagating chains) may be used to couple two growing chains.[44] When polymerization was initiated by a protected hydroxyl (**14**) or carboxyl (**9**) initiator, the resulting telechelic α,ω-diol (**67**) or α,ω-diacid (**68**) would be produced. The best coupling agent for the silyl ketene acetal ended polymer appears to be terephthaloyl fluoride.[125]

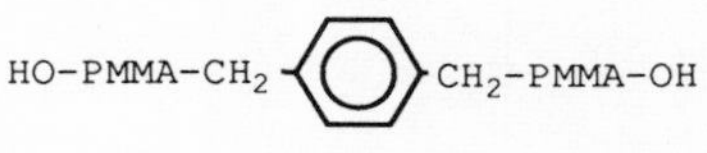

HO-PMMA-CH$_2$-⟨C$_6$H$_4$⟩-CH$_2$-PMMA-OH   HO$_2$C-PMMA-CH$_2$-⟨C$_6$H$_4$⟩-CH$_2$-PMMA-CO$_2$H

(**67**)    (**68**)

In addition to introducing terminal styryl residues via initiation, Asami *et al.*[41] have introduced the same residue in termination by reaction with (**69**). Telechelic poly(methyl acrylate), with terminal hydroxyl groups, has been prepared (see Section 2.3.2).

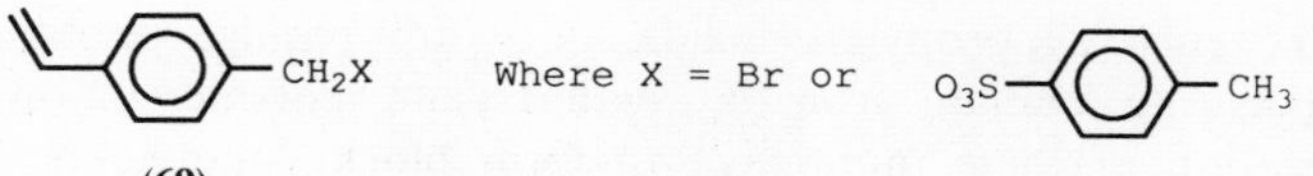

(69) CH=CH-C₆H₄-CH₂X   Where X = Br or O₃S-C₆H₄-CH₃

In principle, other functional terminal groups can be incorporated so long as they can withstand, or be protected against, reaction with the silyl ketene acetal of the propagating species of GTP. There do not appear to be any reports of the synthesis of telechelics through aldol GTP. In principle, any of several groups might be introduced through reaction at the terminal aldehyde group.

2.7 Related and anionic polymerizations

GTP offers cleaner polymerizations of acrylates and methacrylates at ambient temperatures than do conventional anionic polymerizations (Section 2.1). However, GTP propagates by repeated Michael addition and so, in view of developments in anionic polymerization, it is asked 'Is conventional GTP, and possibly other polymerizations, an anionic or modified anionic polymerization?' The propagating species in anionic polymerization of these monomers is an enolate anion, with associated counterion, which adds monomer without further activation. The silyl ketene acetals in GTP are protected enolates which require activation with a catalyst and, it may be argued, activation creates free enolate, if only transitorily.

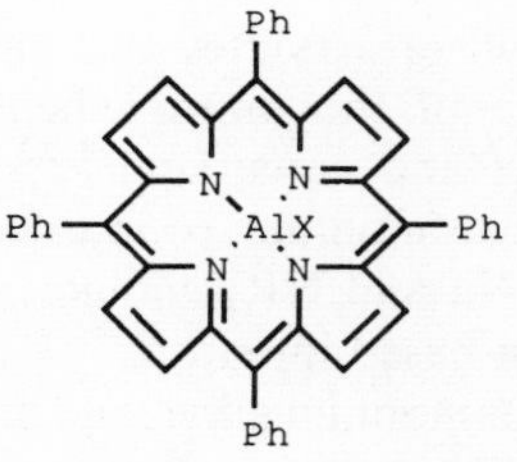

(70) (TPP)AlX

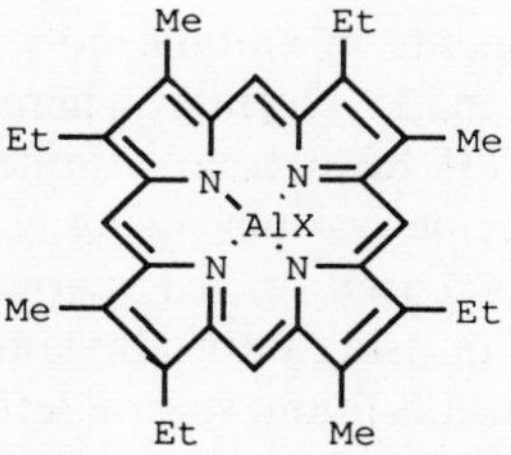

(71) (EtioP)AlX

Aida, Inoue and coworkers reported that certain aluminium porphyrins initiate, quantitatively, living ring-opening polymerizations.[23](See also Section 1.3.2.) They recently reported that similar compounds ((70) with X = Me, Et or CH$_2$Ph; (71) with X = Me) initiate acrylate polymerizations to give polymers with low polydispersity;[20,132] polymerizations are said to involve porphinatoaluminium enolates (72) as the propagating species. Before a methacrylate can add, in either initiation or propagation, activation is required and it is reported that this can be achieved photochemically with visible light ($\lambda > 423$ nm). Thus, propagation involves a photoassisted transfer of the porphinatoaluminium group. Propagation with acrylate and cyclic monomers, such as propylene oxide, does not require photoactivation; the propagating enolate from a methacrylate polymerization initiates polymerization of these monomers to form block copolymers. Thiolates (**70** with X = SR) also initiate without irradiation. Although species (**70**) with X = Cl or OR will initiate polymerization of cyclic monomers they will not initiate polymerization of methacrylates even when irradiated. These several reactions have features in common with silyl ketene acetal-based GTPs and may well be classified as GTPs under a generalization of that title.

$$\text{Me-PPMA} \diagup\hspace{-6pt}=\hspace{-6pt}\diagdown \begin{matrix}\text{OAl(TPP)}\\ \text{OMe}\end{matrix}$$

(**72**)

On the basis of kinetic and molecular weight data from the polymerization of methacrylates, Inoue states that the exchange in equation 2.57 proceeds rapidly, suggesting that the porphinatoaluminium group may become detached during reaction, possibly liberating a free enolate.

$$\text{Polymer}^1 \diagup\hspace{-4pt}=\hspace{-4pt}\diagdown \begin{matrix}\text{OAl(TPP)}\\\text{OR}\end{matrix} \quad + \quad \begin{matrix}\text{(EtioP)AlO}\\\text{RO}\end{matrix} \diagdown\hspace{-4pt}=\hspace{-4pt}\diagup \text{Polymer}^2$$

$$\updownarrow \qquad (2.57)$$

$$\text{Polymer}^1 \diagup\hspace{-4pt}=\hspace{-4pt}\diagdown \begin{matrix}\text{OAl(EtioP)}\\\text{OR}\end{matrix} \quad + \quad \begin{matrix}\text{(TPP)AlO}\\\text{RO}\end{matrix} \diagdown\hspace{-4pt}=\hspace{-4pt}\diagup \text{Polymer}^2$$

Developments in anionic polymerization of acrylates and methacrylates have been made recently whereby some of the improvements achieved through GTP have been attained. Reetz and coworkers[21,22,133] studied anionic polymerizations using tetra-alkylammonium counterions. Against expectations, based on reactivities of Li, Na and NR_4 enolates, they found advantages in using tetra-alkylammonium-based initiators.

These workers found several tetrabutylammonium alkyl- and aryl-thiolates ($RS^{\ominus}{}^{\oplus}NBu_4^n$; R = Bu^n, Ph or $CH_2CH_2OSiMe_3$) to be effective initiators for metal-free anionic polymerizations of acrylates at room temperature in the absence of added catalysts.[21,133] These initiators gave polymers ($< 1500 \text{ g mol}^{-1}$) with low polydispersity and good correlations between values of $\bar{M}_{n,\text{obs}}$ and $\bar{M}_{n,\text{calc}}$. A chain transfer process restricts molecular weight; results obtained depended critically on the nature of the solvent. In spite of the existence of chain transfer, reaction mixtures remain active for some time after polymerization. Poly(methyl acrylate) produced by this process is isotactic, in contrast to the atactic polymers produced by GTP. Reaction proceeds through enolate intermediates (**73**). It was suggested that strong interactions between the enolate and tetra-alkylammonium ions influence polymer tacticity and reduce the rate of termination.

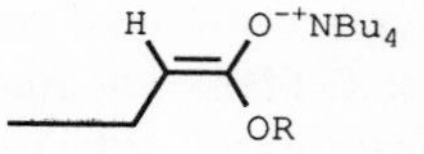

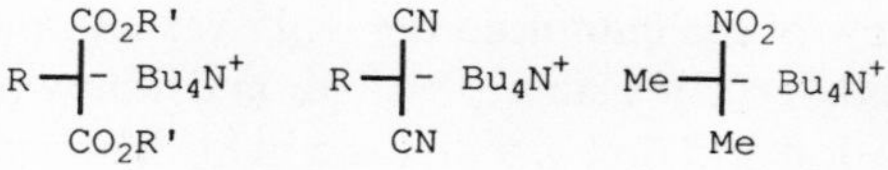

(73)

(74)

R = R' = Et; R = Et, R' = Me

Subsequently, resonance-stabilized metal-free ammonium methanides (74) were found to be good initiators for the living anionic polymerizations of acrylates in tetrahydrofuran or toluene at room temperature;[21,22] butyl methacrylate can be polymerized in this way.[22] The corresponding sodium salts are not effective initiators. Reasonable molecular-weight control, inferior to the best achieved using thiolate initiators, is retained to higher molecular weight ($\sim 20 \text{ kg mol}^{-1}$). The difunctional initiator (75) has also given rise to living polymerizations. It was speculated that strong electrostatic interactions between enolate and counter ions are involved. The positive charge on the tetra-alkylammonium ion resides primarily on the C atom α to the N and long-chain (octyl) alkyl equivalents to (74) were not found to be good initiators,[21] presumably because of the separation of enolate and counterion. It was speculated that the absence of a metal ion to activate the ester function and aid elimination of methoxide reduces the possibility of termination reactions.

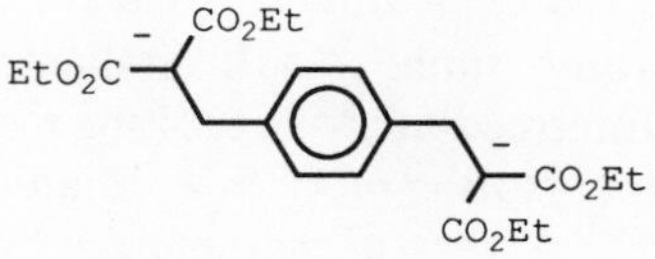

(75)

In another development Fayt et al.[134] modified the environment of the propagating ion pair and reduced the tendencies for chain transfer or termination. Addition of lithium chloride to anionic polymerizations of acrylates with bulky ester groups gave more controlled living polymerizations. Thus, the propagating anion of α-methylstyrene, in the presence of added lithium chloride, initiates polymerization of t-butyl acrylate in tetrahydrofuran/toluene mixtures at $-78\,°\text{C}$ to give a living anionic polymerization.

Pyridine will modify anionic polymerizations of N,N-dimethylacrylamide, ε-caprolactone and alkyl methacrylates leading, in the latter case, to monodisperse PMMA at temperatures greater than normal for anionic polymerization.[135] These several studies, along with GTPs, indicate that possibilities exist for overcoming the inherent problems of anionic polymerization of polar monomers. The object is to eliminate termination reactions, so prevalent with 'free' enolate groups and still present with silyl ketene acetals in the presence of catalyst.

In view of the enhanced control over anionic polymerizations of acrylates and methacrylates now possible, and knowing that GTP is not free from termination and that termination in both processes leads to the same product, the distinction between anionic and group transfer polymerizations has become blurred. Does the use of bulky counterions, or the addition of LiCl, in anionic polymerization modify the environment of the enolate anion so that, because of inherent reactivity, electrostatic interactions or steric effects, the probability of termination relative to propagation is reduced? Does intervention of the catalyst in GTP generate a transient enolate? Are enolate anions formed in modified anionic polymerizations? Is there a true distinction between anionic and group transfer polymerizations? Information on the detailed natures of propagating species and transition states is required to provide answers to the first three questions. In the absence of this information, the fourth remains somewhat philosophical.

A practical distinction between GTP and anionic polymerization of methacrylates remains, although processes other than those based on silyl ketene acetals might be placed under the general heading of GTP. Factors which provide this distinction have been discussed but may be summarized.

In GTP a stable covalent species is activated by a catalyst (or light[20,132]); a brief report refers to the uncatalysed polymerization of MMA initiated by a titanium enolate, presumably through a group transfer process, to living polymers with $D \simeq 1.4$[52]. In GTP only a small fraction of the propagating chains are active at any time. NMR indicates the presence of a pentacoordinate silicon intermediate,[82] suggesting that a true enolate may not be formed during reaction. Ratios of k_p to k_t in anionic and group transfer polymerizations differ.[90,97] Arrhenius parameters have been used to argue for a similarity between anionic and group transfer polymerizations.[95,96] Tacticities of PMMA produced by GTP (atactic) are similar to those from conventional anionic polymerization with well-solvated counterions and differ from those with closely associated counterions (isotactic).[96,97]

Obviously possibilities exist for modifying the nature of propagating species and the structures of the transition states for the several reactions discussed. Considerable work is required to elucidate the natures of the transition states involved and the factors which control the nature and concentrations of active species.

2.8 Applications

GTP provides a superior and more convenient means of synthesizing acrylate and methacrylate polymers with controlled molecular weight and molecular weight distribution than was hitherto possible. The technique, however, has limitations because of the inherent termination reaction, which ultimately limits the lifetime of a propagating species and because of the sensitivity of the silyl ketene acetals to impurities. The technique is best suited to the synthesis of low molecular weight polymers and GTP will not compete with conventional free-radical polymerization for the bulk synthesis of high molecular weight PMMA.

In addition, GTP initiators and PMMA prepared by GTP are expensive. Chain transfer reduces the cost per polymer molecule but reduces the product molecular weight. We expect GTP to find application in specialized situations, where the polymer is valuable or is used in small quantities.

Mention has been made of the use of block copolymers as dispersants for use where one block, say PMMA, is soluble in a matrix and the other block, e.g. ring-opened poly(glycidyl methacrylate), is adsorbed onto a filler or pigment particle.[71] Another use for GTP is the synthesis of copolymers in which one component is derived from a difunctional monomer and only one functionality is used in the GTP, i.e. glycidyl and allyl methacrylates. The glycidyl or allyl group is then used for subsequent modifications of the polymer chains.

Polymers with reactive end groups are potential building blocks for the construction of larger structures. Thus hydroxy- and carboxy-ended polymers may be made through the use of initiators with protected functionalities, as discussed. Other possibilities were also described above. Comb polymers, in which the backbone or the teeth are prepared by GTP, may also find application.

A variety of star-shaped molecules can be synthesized. Simple, soluble star polymers with n arms may be prepared by the addition monomer to n-functional initiator prepared in situ.[136] Star polymers may also be prepared by aldol-GTP through the use of multifunctional initiators.[86,111]

More complex molecular structures are available by use of GTP. Addition of a difunctional monomer, e.g. ethylene glycol dimethacrylate, to living PMMA generates blocks which cross-link during polymerization. This cross-linked core has many arms attached and is a star-like structure with PMMA blocks attached to the surface. Reference has been made to the use of these stars for production of a tough plastic coating.[69] If, in making the multi-armed stars by GTP, the initiator has a protected hydroxyl then, after deprotection, the star is a polyol capable of being linked to other structures to form highly cross-linked materials.[71]

A novel application of GTP is in the synthesis of ladder polymers. It is reported that polymerization of a difunctional monomer possessing long spacers between the functionalities, e.g. hexamethylene glycol dimethacrylate, with a difunctional initiator in dilute solution does not give a cross-linked, nor cyclopolymerized polymer. Rather a soluble product, thought to be a ladder polymer, is formed.[47] By adding a conventional difunctional monomer to the living ladder polymer, a multi-ladder-armed star with a cross-linked core can be prepared.[47]

References

1. Cowie, J.M.G., in *Comprehensive Polymer Science*, (Eds. Eastmond, G.C., Ledwith, A., Russo, S. and Sigwalt, P.), Pergamon Press, Oxford, Vol. 4, p. 1 (1988).
2. Fontanille, M., in *Comprehensive Polymer Science*, (Eds. Eastmond, G.C., Ledwith, A., Russo, S. and Sigwalt, P.), Pergamon Press, Oxford, Vol. 4, p. 365 (1988).
3. Szwarc, M., Levy, M. and Milkovich, R. *J. Amer. Chem. Soc.* **78**, 2656 (1956); Szwarc, M., *Nature* **178**, 1168 (1956).
4. Worsfold, D.J. and Bywater, S. *Can. J. Chem.* **38**, 1891 (1960).
5. Yuki, H., Hatada, K., Niinomi, T. and Kikuchi, K., *Polymer J.* **1**, 36 (1970).
6. Müller A.H.E. in *Recent Advances in Anionic Polymerization*, (Eds. Hogen-Esch, T.E. and Smid, J.), Elsevier, New York, p. 205 (1987).
7. Croucher, T.G. and Wetton, R.E., *Polymer* **17**, 205 (1976).
8. Miyamoto, M., Sawamoto, M. and Higashimura, T. *Macromolecules* **17**, 265 (1984).
9. Higashimura, T. and Miyamoto, M. *Macromolecules* **18**, 611 (1988).
10. Enoki, T., Swamoto, M. and Higashimura, T. *J. Polym. Sci., Polym. Chem. Ed.* **24**, 2261 (1986).
11. Higashimura, T., Aoshima, S. and Swamoto, M. *Makromol. Chem., Macromol. Symp.* **13/14**, 457 (1988).
12. Faust, R. and Kennedy, J.P. *J. Polym. Sci., Polym. Chem. Ed.* **25**, 1847 (1987).
13. Szwarc, M. *Advan. Polym. Sci.* **49**, 133 (1983).
14. Hatada, K., Kitayama, T., Fumikawa, K., Ohta, K. and Yuki, H. *ACS Symp. Ser.* **166**, 327 (1981).
15. Müller, A.H.E., Lochmann, L. and Trekoval, J. *Makromol. Chem.* **187**, 1473 (1986).
16. Lochmann, L. and Müller, A.H.E. Preprints of IUPAC International Symposium on Macromolecules, Merseburg (1987), p. 75.
17. Webster, O.W., Hertler, W.R., Sogah, D.Y., Farnham, W.B. and RajanBabu, T.V. *J. Amer. Chem. Soc.* **105**, 5706 (1983).
18. Webster, O.W., Farnham, W.B. and Sogah, D.Y. European Patent Application 0 068 887 (1983).
19. Webster, O.W. US Patent 4,417,034 (1983).
20. Inoue, S. *Polym. Prepr.* **29**(2), 42 (1988).
21. Trofimoff, L., Aida, T. and Inoue, S. *Chem. Lett.*, 991 (1987) and earlier papers.
22. Reetz, M.T., *Angew. Chem. Int. Ed. Engl.* **27**, 994 (1988).
23. Reetz, M.T., Knauf, T., Minet, U. and Bingel, C. *Angew. Chem. Int. Ed. Engl.* **27**, 1373 (1988).
24. Colvin, E.W. *Chem. Soc. Rev.* **7**, 15 (1978).
25. Wiles, D.M. and Bywater, S. *J. Polym. Sci.* **2**(B) 1175 (1964).
26. Andrews, G.D. and Melby, L.R. in *New Monomers and Polymers* (Ed. Culbertson, B.M.), Plenum, New York, p. 357 (1984).
27. Brownbridge, P. *Synthesis* **1**, 85 (1983).
28. Mukaiyama, T. *Angew. Chem.* **89**, 858 (1977); *Angew. Chem. Int. Ed. Engl.* **16**, 817 (1977).
29. Colvin, E.W. *Silicon in Organic Synthesis*, Butterworth, London (1981).
30. Colvin, E.W. Silicon Reagents in Organic Synthesis, in *Best Synthetic Methods*, Academic Press, London (1988).
31. Fleming, I. *Comprehensive Organic Chemistry* (Ed. Jones, D.N.), Pergamon Press, Oxford Vol. 3, p. 541 (1979).
32. Fleming, I. *Chem. Soc. Rev.* **10**, 83 (1981).
33. Fleming, I. *Chimia*, 265 (1980).
34. Rasmussen, J.K. *Synthesis* 91 (1977).
35. Ainsworth, C. and Kuo, Y.-N. *J. Organometallic. Chem.* **46**, 73 (1972).
36. Adam, W. and Cueto, O. *J. Org. Chem.* **42**, 38 (1977).
37. Ainsworth, C., Chen, F. and Kuo, Y.-N. *J. Organometallic Chem.* **46**, 59 (1972).
38. Kuo, Y.-N., Chen, F., Ainsworth, C. and Bloomfield, J.J. *Chem. Commun.* 136 (1971).
39. Bandermann, F., Sitz, H.-D. and Speikamp, H.-D. *Polym. Prepr.* **27**(1), 169 (1986).
40. Sogah, D.Y., Hertler, W.R., Webster, O.W. and Cohen, G.M. *Macromolecules* **20**, 1473 (1987).
41. Asami, R., Kondo, Y. and Takaki, M. *Polym. Prepr.* **27**(1), 186 (1986).
42. Kuwajima, I. and Nakamura, E. *J. Amer. Chem. Soc.* **90**, 4464 (1968).

43. Andrews, G.D. and Vatvars, A. *Macromolecules* **14**, 1603 (1981).
44. Sogah, D.Y. and Webster, O.W. *J. Polym. Sci. Lett. Ed.* **21**, 927 (1983).
45. Webster, O.W., Hertler, W.R., Sogah, D.Y., Farnham, W.B. and RajanBabu, T.V. *J. Macromol. Sci., Chem.* **A21**, 943 (1984).
46. Eastmond, G.C., Gilchrist, T.L., Lau, J. and Page, P.C.B. unpublished results.
47. Sogah, D.Y. *Polym. Prepr.* **29**(2), 3 (1988).
48. Yu, H.-S., Choi, W.-J., Lim, K.-T. and Choi, S.-K. *Macromolecules* **21**, 2893 (1988).
49. Hertler, W.R., RajanBabu, T.V., Ovenall, D.W., Reddy, G.S. and Sogah, D.Y. *J. Amer. Chem. Soc.* **110**, 5841 (1988).
50. Hertler, W.R., RajanBabu, T.V. and Sogah, D.Y. *Polym. Prepr.* **29**(2), 71 (1988).
51. Bandermann, F. and Speikamp, H.D. *Makromol. Chem., Rapid Commun.* **6**, 335 (1985).
52. Reetz, M.T. *Pure and Appl. Chem.* **67**, 1785 (1985).
53. Webster, O.W., U.S. Patent 4,508,880 1985.
54. Farnham, W.B., Sogah, D.Y. U.S. Patents 4,414,372 1983; 4,524,196, 1985; 4,581,428, 1986.
55. Webster, O.W., Farnham, W.B., Sogah, D.Y. EPA 145,263 1984.
56. Reetz, M.T., Ostarek, R. and Pie jko, K-E, German Patent Application 3,504,168, 1985; US Patent 4,626,579 1986.
57. Hertler, W.R. *Polym. Prepr.* **27**(1), 165 (1986).
58. Webster, O.W. US Patent 4,417,034 Nov. 22, 1983.
59. Speikamp, H.-D. and Bandermann, F. *Makromol. Chem.* **189**, 437 (1988).
60. Sitz, H.-D., Speikamp, H.-D. and Bandermann, F. *Makromol. Chem.* **189**, 429 (1988).
61. Dicker, I.B., Cohen, G.M., Farnham, W.B., Hertler, W.R., Laganis, E.D. and Sogah, D.Y. *Polym. Prepr.* **28**(1), 106 (1987).
62. Dicker, I.B., Farnhma, W.B., Hertler, W.R., Laganis, E.D., Sogah, D.Y., del Pesco, T.W., Fitzgerald, P.H. US Patent 4,588,795 1986; CA 86 **105**, 98135.
63. Shneider, L.V. and Dicken, I.B. US Patent 4,736,003 1988.
64. Pickering A. and Thorne, A.J. US Patent 4,791,181 1988.
65. Chou, S.S.P. and Niu, C.W. *MRL Bull. Res. Dev.* **29**, 808 (1987).
66. Kawai, M., Onaka, M. and Izumi, Y. *Bull. Chem. Soc. Japan.* **61**, 2157 (1988).
67. Kreuder, W., Webster, O.W. and Ringsdorf, H. *Makromol. Chem., Rapid Commun.* **7**, 5 (1986).
68. Pugh, C. and Percec, V. *Polym. Bull.* **14**, 109 (1985).
69. Sogah, D.Y., Hertler, W.R. and Webster, O.W. *Polym. Preprints, Polym. Div., Amer. Chem. Soc.* **25**(2), 3 (1984).
70. Hertler, W.R., Sogah, D.Y., Webster, O.W. and Trost, B.M. *Macromolecules* **17**, 1415 (1984).
71. Simms, J.A. and Spinelli, H.J. *J. Coat. Technol.* **57**, 125 (1987).
72. Gomez, P.M. and Neidlinger, N.H. *Polym. Prepr.* **29**(1), 209 (1987).
73. Butler, G. in *Comprehensive Polymer Science* (Eds. Eastmond, G.C., Ledwith, A. Russo, S. and Sigwalt, P.), Vol. 4, p. 1 Pergamon Press, Oxford, (1988).
74. Quirk, R.P. and Bidinger, G.P. *Polym. Preprints. Polym. Div. Amer. Chem. Soc.* **29**(2), 120 (1988).
75. Müller, A.H.E. and Stickler, M. *Makromol. Chem., Rapid Commun.* **7**, 575 (1986).
76. Wei, Y. and Wnek, G.E. *Polym. Prepr.* **28**(1), 252 (1987).
77. Brittain, W.J., Davidson, F. and Reddy, G.S. paper presented at 8th International Symposium on Organosilicon Chemistry, St. Louis 1987.
78. Brittain, W.J. *Polym. Prepr.* **29**(2), 312 (1988).
79. Akkaped, M.K. *Macromolecules* **12**, 546 (1979).
80. House, H.O. in *Modern Synthetic Reactions*, W.A. Benjamin, Philippines (1972).
81. Sogah, D.Y. and Farnham, W.B. in *Organosillicon and Bioorganosilicon Chemistry: Structure, Bonding, Reactivity and Synthetic Applications* (Ed. Sakurai, H.), Ch. 20, J. Wiley and Sons, N.Y. (1985).
82. Farnham, W.B. and Sogah, D.Y. *Polym. Prepr.* **27**(1), 167 (1982).
83. Farnham, W.B. and Harlow, R.L. *J. Amer. Chem. Soc.* **103**, 4608 (1981); Farnham, W.B. and Whitney, J.F. *J. Amer. Chem. Soc.* **106**, 3392 (1984); Perkins, C.W., Martin, J.C., Arduengo, A.J., Lau, W., Alegria, A. and Kochi, J. K. *J. Amer. Chem. Soc.* **102**, 7753 (1980); Michalak, R.S. and Martin, J.C. *J. Amer. Chem. Soc.* **104**, 1683 (1982); Font, J.J.H.M., Freide, W. and Trippett, S. *J. Chem. Soc., Chem. Commun.* 934 (1980); Totsch, W. and Sladky, F. *J. Chem. Soc., Chem. Comum.* 927 (1980).
84. Dixon, D.A., Sogah, D.Y. and Farnham, W.B. unpublished results, quoted in Ref. 86.
85. Dixon, D.A. unpublished results, quoted in Ref. 86.

86. Sogah, D.Y. and Webster, O.W. in 'Recent Advances in Mechanistic and Synthetic Polymerization' (Eds. M. Fontanille and A. Guyot), *NATO ASI Ser C* **215**, Reidel, Dordrecht p. 61 (1987).
87. Brittain, W.J. *J. Amer. Chem. Soc.* **110**, 7440 (1988).
88. Gennick, I., Harmon, K.M. and Potvin, M.M. *Inorg. Chem.* **16**, 2033 (1977); Fujiwara, F.Y., Martin, J.S. *J. Amer. Chem. Soc.* **96**, 7625 (1974).
89. Dicker, I.B., Cohen, G.M., Farnham, W.B., Hertler, W.R., Laganis, E.D. and Sogah, D.Y. *Macromolecules* (in press).
90. Brittain, W.J. and Dicker, I.B. *Macromolecules* **22**, 1054 (1989).
91. Hertler, W.R. *Macromolecules* **20**, 2976 (1987).
92. Hertler, W.R. *Polym. Prepr.* **28**(1), 108 (1987).
93. Hertler, W.R., Dixon, D.A., Matthews, E.W., Davidson, F. and Kitson, F.G. *J. Amer. Chem. Soc.* **109**, 6532 (1987).
94. Mai, P.M. and Müller, A.H.E. *Makromol. Chem., Rapid Commun.* **8**, 99 (1987).
95. Mai, P.M. and Müller, A.H.E. *Makromol. Chem., Rapid Commun.* **8**, 247 (1987).
96. Müller, A.H.E. in 'Recent Advances in Mechanistic and Synthetic Aspects of Polymerization' *NATO ASI Ser C*, (Eds. Fontanille, M. and Guyot, A.), **215**, D. Reidel, Dordrecht, Ref. p. 23 (1987).
97. Doherty, M.A. and Müller, A.H.E. quoted in Ref. 95.
98. Doherty, M.A., Gores, F., Mai, P.M. and Müller, A.H.E. *Polym. Prep.* **29**(2), 73 (1988).
99. Noyori et al. *J. Amer. Chem. Soc.* **102**, 1223 (1980).
100. Sitz, H.-D. and Bandermann, F. in 'Recent Advances in Mechanistic Aspects of Polymerization', *NATO ASI Ser C*, (Eds. Fontanille, M. and Guyot, A.), **215**, D. Reidel, Dordrecht, p. 41, (1987).
101. Busfield, W.K. and Methven, J.M. *Polymer* **14**, 137 (1973).
102. Dicker, I.B. *Polym. Prepr.* **29**(2), 114 (1988).
103. Bevington, J.C. and Ebdon, J.R., in (Ed. R.N. Haward), *Developments in Polymerization-2*, Applied Science, London, p. 26 (1979).
104. Kabanov, V.A. *Makromol. Chem., Macromol. Symp.* **10/11** 193 (1987).
105. Bamford, C.H., in *Comprehensive Polymer Science*, (Eds. Eastmond, G.C., Ledwith, A., Russo, S. and Sigwalt, P.), (1988). Vol. 4, p. 219, Pergamon Press, Oxford.
106. Paterson, I. and Fleming, I. *Tetrahedron Lett.* 993 (1979).
107. Reetz, M.T., Maier, W.F., Schwellmis, K. and Chatsohosifidis, I. *Angew. Chem. Int. Ed. Engl.* **18**, 72 (1979).
108. Kita, Y., Segawa, J., Haruta, J., Fujii, T., Kitagawa, Y., Yamamoto, H. and Nozaki, H. *J. Amer. Chem. Soc.* **99**, 4192 (1977).
109. Dicker, I.B. Abstracts of 192nd ACS National Meeting (1986).
110. Sogah, D.Y. *Polym. Prep.* **27**(1), 163 (1986).
111. Sogah, D.Y. and Webster, O.W. *Macromolecules* **19**, 1775 (1986).
112. Fleming, I. *Chimia* 265 (1980).
113. Murahashi, S., Nozakura, S. and Sumi, M. *J. Polym. Sci. Pt. B* **3**, 245 (1965).
114. Nozakura, S., Ishihara, S., Inaba, Y. and Matsumura, K. *J. Polym. Sci., Pt A-1*, **11**, 1053 (1973).
115. Solara, R. and Chiellini, E. *Gazz. Chim. Ital.*, **U106**, 1037 (1976).
116. Corey, E.J. and Snider, B.B. *J. Amer. Chem. Soc.* **94**, 2549 (1972).
117. Ackerman, E. *Acta Chem. Scand.* **11**, 373 (1957).
118. Hirabayashi, T., Itoh, T. and Yokota, K. *Polymer J.* **20**, 1041 (1988).
119. Jenkins, A.D., Tsartolia, E., Walton, D.R.M., Stejskal, J. and Kratochvil, P. *Polymer Bull.*, **20**, 97 (1988).
120. Bevington, J.C. and Harris, D.O. *J. Polym. Sci.* **B5**, 799 (1976).
121. Musha, Y., Hori, Y., Sato, Y. and Katayama, M. *Nihon Daigaku Kogakubu Kiyo, Bunri* **A26**, 175 (1985).
122. Pugh, C. and Percec, V. *Polym. Prepr.* **26**(2), 303 (1985).
123. Hutchins, C.S. and Shor, A.C., US Patent 4,656,226 (to E.I. du Pont de Nemours & Co. Inc.) (1987).
124. Eastmond, G.C. and Grigor, J. *Makromol. Chem., Rapid Commun.* **7**, 375 (1986).
125. Cohen, G.M. *Polym. Prepr.* **29**(2), 46 (1988).
126. Hellerstein, A.M., Smith, S.D. and McGrath, J.E. *Polym. Prepr.* **28**(2), 328 (1987).
127. Hellerstein, A.M., DeSimone, J.M. and McGrath, J.E. *Polym. Prepr.* **29**(1), 148 (1988).

128. Risse, W. and Grubbs, R.H. *Macromolecules* **22**, 1558 (1989).
129. Cannizo, L.F. and Grubbs, R.H. *J. Org. Chem.* **50**, 2316 (1985); Clawson, L.E., Buchwald, S.L. and Grubbs, R.H. *Tetrahedron Lett.*, 5733 (1984); Brown-Wensley, K.A., Buchwald, S.L., Cannizo, L.F., Ho, S., Meinhardt, D., Stille, J.R., Straus, D. and Grubbs, R.H. *J. Amer. Chem. Soc.* **102**, 3270 (1980).
130. March, J. *Advances in Organic Chemistry: Reaction Mechanisms and Structures*, Wiley, NY (1985).
131. Jung, M.E. and Blum, R.B. *Tetrahedron Lett.*, 3791 (1977).
132. Kuroki, M., Nashimoto, S., Aida, T. and Inoue, S. *Macromolecules* **21**, 3114 (1988).
133. Reetz, M.T. and Ostarek, R. *J. Chem. Soc., Chem. Commun.*, 213 (1988).
134. Fayt, R., Forte, R., Jacobs, C., Ouhadi, T., Teyssie, Ph. and Varshney, S.K. *Macromolecules* **20**, 1442 (1987).
135. Hunh-Ba, G. and McGrath, J.E., in (Eds. T.E. Hogen-Esch and J. Smid), *Recent Advances in Anionic Polymerization*, Elsevier, New York, p. 173 (1987).
136. Webster, O.W. and Sogah, D.Y. in (Eds. Fontanille, M. and Guyot, A.) *Recent Advances in Mechanistic and Synthetic Polymerization*, NATO ASI Ser. C 215 Reidel Pub. Corp., Dordrecht, p. 3 (1987).

3 Ring-opening metathesis polymerization of cyclic alkenes

A.J. AMASS

The ring-opening polymerization of cyclic alkenes is a special case of the more general olefin metathesis reaction. A number of specialist reviews[1] have appeared in the literature on both the metathesis of alkenes and ring-opening polymerization. The metathesis reaction was initially discovered by Banks and Bailey[2] but it was Calderon[3] who first recognized the relationship between ring-opening polymerization and the olefin metathesis, or disproportionation, reaction. The reaction is catalysed by complexes derived principally from group VI metal compounds but group IV and V compounds have also been used. Effective catalysts may be either homogeneous or heterogeneous and, although not essential as part of the catalyst system, organometallic compounds, usually either organometallic aluminium or tin compounds, are incorporated as a second component of two component catalyst systems.

3.1 Scope of ring-opening metathesis polymerization

Olefin metathesis may be summarized by equation 3.1.

$$2 R^1CH=CHR^2 \rightleftharpoons R^1CH=CHR^1 + R^2CH=CHR^2 \quad (3.1)$$

The carbon–carbon double bond may be part of either a linear alkene or a cyclic system. If the olefin is cyclic then the result of the metathesis reaction is polymerization, as shown in equation 3.2

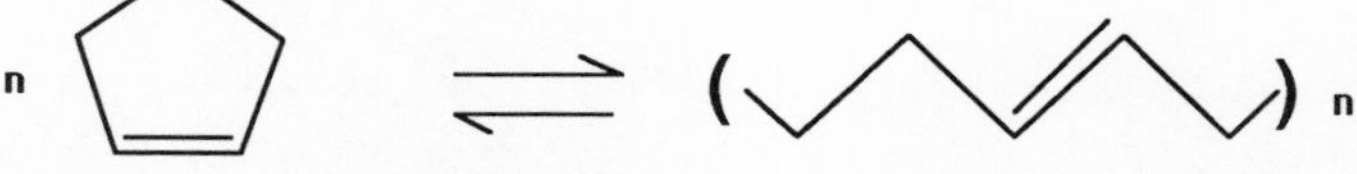

$$n \; \bigcirc \;\rightleftharpoons\; (\sim\!\sim\!\sim)_n \quad (3.2)$$

The olefin metathesis reaction is an equilibrium reaction and, like all true equilibrium reactions, the position of equilibrium can be approached from either side of the equation. Thus, in the above olefin metathesis equation 3.1, metathesis of equal numbers of moles of the product olefins yields an identical mixture at equilibrium. Olefin metathesis is a redistribution reaction in which the reactants and products have the same reactive functionality. Cross

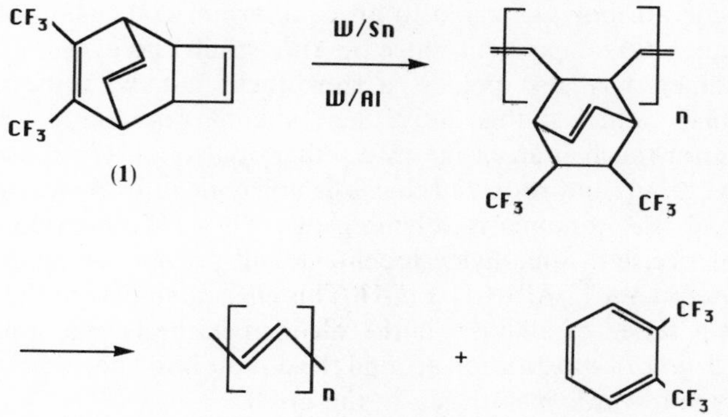

Figure 3.1

metathesis can be carried out not only between linear olefins as shown above but also between a cyclic and a linear olefin, in which case the linear olefin effectively acts as a transfer agent during the polymerization reaction and may be used to control the molecular weight of the polymer chain. The structure of the polymer formed by the cross metathesis of a cyclic and linear alkene would be as indicated in equation 3.3.

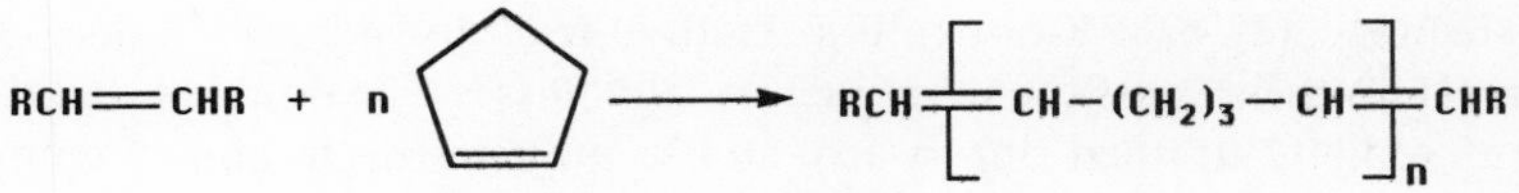

$$RCH=CHR + n\,\bigcirc \longrightarrow [RCH=CH-(CH_2)_3-CH=CHR]_n \quad (3.3)$$

Ring-opening metathesis polymerization may be used to produce a range of polymers and copolymers and in some cases the products of the polymerization may be used as precursors in subsequent reactions. The polymerization of 7,8-bis-trifluoromethyltricyclo-[4.2.2.0^{2,4}]-deca-3,7,9-triene[4] (1) leads to a polymer from which hexafluoroxylene can be eliminated to form polyacetylene, as shown in Figure 3.1.

3.2 Monomers for ring-opening metathesis polymerization

The ring-opening polymerization of cycloalkenes takes place provided that the carbon–carbon double bond is part of a strained ring system. Of the simple cycloalkenes, cyclohexene does not undergo ring-opening metathesis polymerization because the monomer is strain-free, although there have been reports by McCarthy[5] that ring strain energy is not an important factor in determining the polymerizability of a monomer. Cyclooctene, cyclobutene and 1,5-cyclooctadiene are all polymerizable to high, almost 100%, conversion, being all highly sterically strained monomers. Cyclopentene and cycloheptene are polymerizable to an equilibrium conversion of monomer that is temperature dependent, since the ring-strain energy is only moderate. The monomer may also possess a substituent but, with the majority of conventional catalysts, this substituent should not contain an electron donating group such as an amine, ester, ether or alcohol. The substituent may be an alkyl group and provided that it is not bonded to the carbon–carbon double bond, the monomer is polymerizable. Thus 1-methycyclopentene will not polymerize, but 3-methylcyclopentene will polymerize readily with the soluble catalyst $WCl_6/AlEtCl_2/EtOH$. This effect is similar to the reactivities found for a series of alkenes in the olefin metathesis reaction, when the increased degree of substitution around the double bond decreases the activity of the alkene towards metathesis, in the order:

$$(R)_2C\!=\!C(R')_2 < RCH\!=\!C(R')_2 \ll RCH\!=\!CHR' < RCH\!=\!CH_2$$

The polymer obtained on the ring-opening polymerization of a monocyclic olefin is effectively only a high molecular weight olefin and, as such, the carbon–carbon double bonds of the polymer may participate in further, secondary, metathesis reactions. Changes in the stereochemistry and the molecular weight distribution of the first formed polymer often result from such secondary reactions.

The monomer may also be a bicyclic or tricyclic alkene, such as bicyclo-2.2.1-heptene (**2**) or 7,8-bis(trifluoromethyl)-tricyclo-[$4.2.2.0^{2.5}$]-deca-3,7,9-triene (**1**). In these cases the monomer usually possesses a double bond that is part of a highly strained ring system and so polymerizes to almost complete conversion. Such polymers do not readily participate in secondary metathesis reactions and as a consequence norbornene monomers have been used to investigate the mechanism of stereocontrol of ring-opening polymerization, which will be discussed in another section. The ring-opening polymerization of strained bicyclic monomers is shown in Figure 3.2. In the extension of the principles of ring-opening polymerization to polycyclic monomers the carbon–carbon double bond that undergoes metathesis is that associated with the greatest ring strain energy.

Feast[6] has studied the effects of halogen substituents on the reactivity of a monomer towards the ring-opening polymerization reaction. If the substituent is bonded to the carbon–carbon double bond the monomer is unreactive towards polymerization; furthermore, if the substituent is a fluorine atom the monomer is not polymerizable if the fluorine atom is part of the same ring as the double bond. This is also true when the monomer is bicyclic; if the substituent is part of a ring other than that containing the double bond the monomer will polymerize with ease. Examples of some fluorinated polymers that may be prepared in this way are shown in Figure 3.3.

The metathesis reaction, applying as it does to linear and cyclic alkenes, can be performed on a mixture of the two alkenes, and it is found that the linear alkene acts as a molecular weight control[7] during the polymerization of the

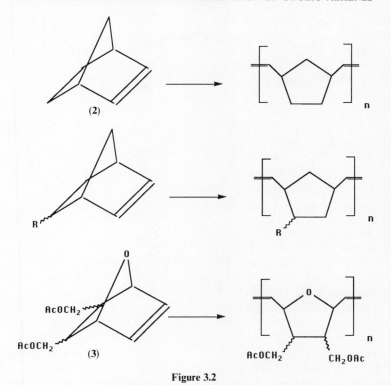

Figure 3.2

cyclic alkene. Similarly if the cyclic alkene is copolymerized with a bicyclic diene, such as norbornadiene,[8] then a highly crosslinked network is obtained; in fact the metathesis polymerization of dicyclopentadiene is the basis of an industrial process by Hercules. The other important industrial process, also based on this type of technology, is the synthesis, by Chemische Werke Huels, of low molecular weight polyoctenylene, Vestenamer®, by metathesis polymerization of cyclooctene with a linear alkene.

Examples of the metathesis polymerization of functionalized monomers are rare, although much work has been carried out on the metathesis of linear unsaturated esters. Unsaturated esters have been metathesized using heterogeneous catalysts[9] but the efficiencies of such reactions are known to be low, because catalyst demands are high. Amass[10] found that the ring-opening metathesis polymerization of cyclopentene may be carried out in the presence of an ester using $WCl_6/AlEtCl_2/EtOH$ as a catalyst, if the ester contains an electron withdrawing group close to the ester group. Thus catalytic proportions of the catalyst were required for the polymerization of cyclopentene with 3-hexenyl-1-trichlorethanoate and low molecular weight products were obtained. The active centre in metathesis is known to be a

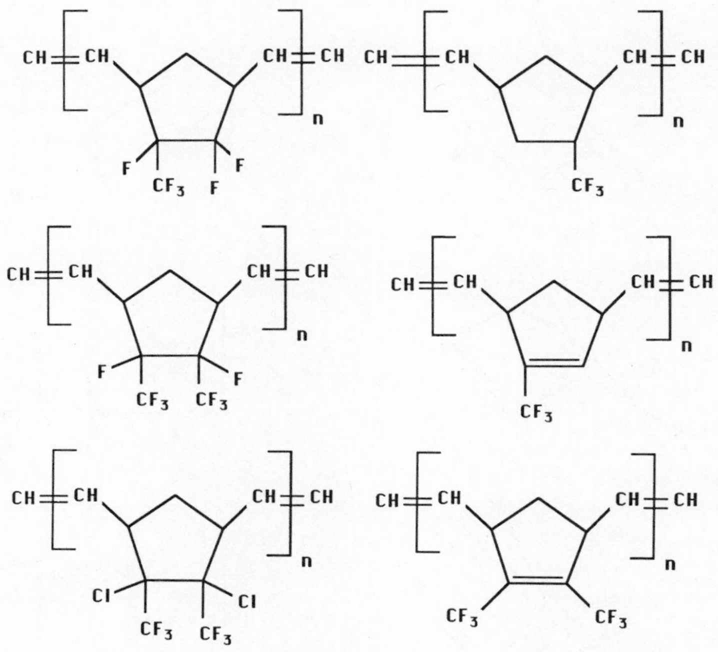

Figure 3.3

transition metal carbene which possesses a vacant coordination site and normally any electron rich compound will preferentially complex with the site and block the propagation centre. Recent reports by Grubbs,[11] however, have suggested that $RuCl_3$ will act as a metathesis catalyst under aqueous conditions and monomers such as ethers, e.g. 5,6-bis(acetoxymethyl)-7-oxabicyclo-[2.2.1]-heptene (**3**), will polymerize readily.

Reports by Schrock[12] suggest that a stable metal carbene may also be used to polymerize the very reactive norbornene monomers that may possess functional groups such as the ester group. The success of this type of polymerization may be associated with the fact that the monomer is very reactive and accordingly, if the reactivity of the double bond is so enhanced in a highly strained system, it can compete effectively with the functional group for the active centre. In addition, the metal carbene is a relatively stable carbene which may also assist the course of the reaction.

3.3 Catalysts for the ring-opening polymerization of cycloalkenes

An extensive range of catalysts has been used to initiate the ring-opening metathesis polymerization of cycloalkenes and to perform alkene metathesis, details of which are to be found in the many reviews and textbooks that have been published on metathesis.[1] The catalysts may be classified broadly as either heterogeneous or homogeneous. The earliest catalysts used for both ring-opening polymerization and alkene metathesis were heterogeneous[2,13] but this type of catalyst is generally unsatisfactory for polymerization reactions and it is for the most part accepted that, for polymerization, catalysts are homogeneous. Suffice it to say here that the heterogeneous catalysts are formed by adsorbing a transition metal compound, normally an oxide or carbonyl compound, such as WO_3, $W(CO)_6$ or Re_2O_7, on to an 'inert' metal oxide support, such as Al_2O_3 or SiO_2. The catalyst is then activated by the interaction with a metal alkyl, e.g. $SnMe_4$ or $AlEtCl_2$.[14]

A one-component catalyst system[15] may be used for ring-opening polymerization and indeed many of the latest catalysts[16,17] are extremely well-defined one-component systems. However, it is usual for a catalyst to comprise at least two, if not more, components. A most general definition of a metathesis catalyst is similar to the definition of a Ziegler catalyst, i.e. a transition metal compound and a metal alkyl. The transition metal compound is normally selected from groups IVA to VIII of the periodic table. Of these metals by far the most commonly used are tungsten, molybdenum and rhenium, although perhaps one should add titanium to this list in view of the recent work published by Grubbs.[18] Other transition metals that have been used include chromium,[19] for alkyne polymerization, niobium,[20] tantalum[21] and ruthenium[22] for alkene and cycloalkene polymerization. The applicability of particular catalyst combinations depends to some extent on the monomer that is to be polymerized or the alkene to be metathesized. For example, the strained ring monomers such as norbornene and its derivatives may be polymerized by catalysts based on such metals as titanium,[18] ruthenium[23] and indeed osmium,[24] whereas the monomers of lower ring strain such as cyclopentene are normally polymerized by the more active catalysts derived from tungsten.

The earliest example of a homogeneous catalyst was WCl_6 activated by $AlEt_3$. Natta[25] described the use of this catalyst combination for the polymerization of cyclopentene, although the yields of polymer were not particularly high. The low activities of catalysts such as those described by Natta are associated with termination reactions that take place during the course of the reaction, which have been shown to occur in such polymerization systems. The efficiency of the polymerization may be considerably improved by using a different cocatalyst; the catalyst system described by Calderon,[26] $WCl_6/AlEtCl_2/EtOH$, produces high yields of polymers with monomers such as cyclopentene, cyclooctene and cyclooctadiene. Other oxygen-containing compounds such as epichlorhydrin,[20] may be used as third component activators for ring-opening polymerization, particularly in the case of monomers such as cyclopentene. Other halides such as $WOCl_4$[27] and WF_6[20] have found use as the catalyst in metathesis but usually these require the use of at least a metal alkyl cocatalyst. Of particular significance in the case of WF_6 is the effect that the molar ratio of the catalyst to the cocatalyst has on the stereochemistry of the polymers produced by the polymerization of cyclopentene. These effects will be considered in a later section.[20]

Some of the earliest one-component catalysts were tungsten compounds, the Fischer and Casey carbenes,[28,29] although they are not normally used to initiate metathesis except if reacted with a cocatalyst, heated, or photochemically promoted. Katz[30] has used the Casey carbene to initiate the polymerization of highly strained monomers such as norbornene but the significant recent development in this area has been the disclosure by Schrock[17] of the initiation of the polymerization of norbornene derivatives by the stable carbene (4).

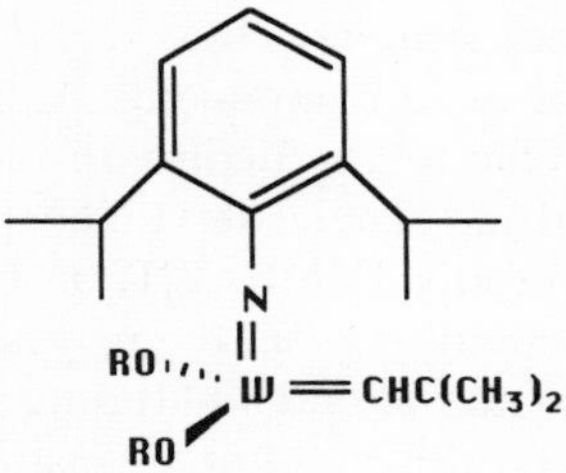

(4)

A considerable amount of work has been carried out on this system and a range of monomers, mainly, but not exclusively, of the norbornene type, has been used to obtain living polymer systems. The rate of polymerization has been used as a measure of the activity of the catalyst and the inductive effect of the substituent group (R) has shown that the electron-withdrawing nature of the group had a significant influence on the rate of polymerization; as the electron-withdrawing effect was increased the rate of polymerization also increased.

Other well-characterized complexes of carbenes have been shown by Ivin et al.[31-34] to be catalysts for the polymerization of norbornene and its derivatives; these catalysts are for example compound (5). The crystal structures of the compound itself and the product of its reaction with $GaBr_3$ have been determined. Furthermore the successive replacement of bromide by neopentoxy ligands has an interesting effect on the rate of metathesis of cis-pent-2-ene in the presence and absence of $GaBr_3$. In the absence of $GaBr_3$, the rate of metathesis was greatest with compound (5) and decreased with increasing replacement of the bromide by neopentoxy groups, whereas when the cocatalyst was incorporated in the catalyst system, the rate of metathesis was two orders of magnitude faster and increased with increasing substitution of the bromide group. The greater rate of polymerization observed when $GaBr_3$ was used was ascribed to the formation of ionic structures, in agreement with the conductometric studies undertaken.

$$Br_2W=CHC(CH_3)_3(OCH_2Bu^t)$$

(5)

Recent work by Grubbs[18] has highlighted the role played by titanium catalysts in the ring-opening polymerization of cyclic alkenes. The catalyst has been derived from Tebbe's reagent (6) prepared by the reaction of titanocene dichloride with aluminium trimethyl, equation 3.4.

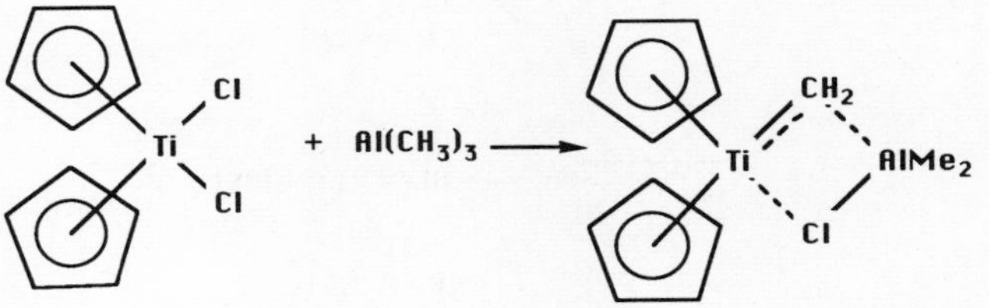

$$Cp_2TiCl_2 + Al(CH_3)_3 \longrightarrow Cp_2Ti(\mu\text{-}CH_2)(\mu\text{-}Cl)AlMe_2 \quad (3.4)$$

(6)

A stable titanacyclobutane may then be obtained by the reaction of the Ti/Al complex with an alkene such as isobutene in the presence of pyridine as a solvent. The resulting metallocyclobutane may be used for the polymerization of strained-ring olefins such as norbornene to living polymers, equation 3.5.

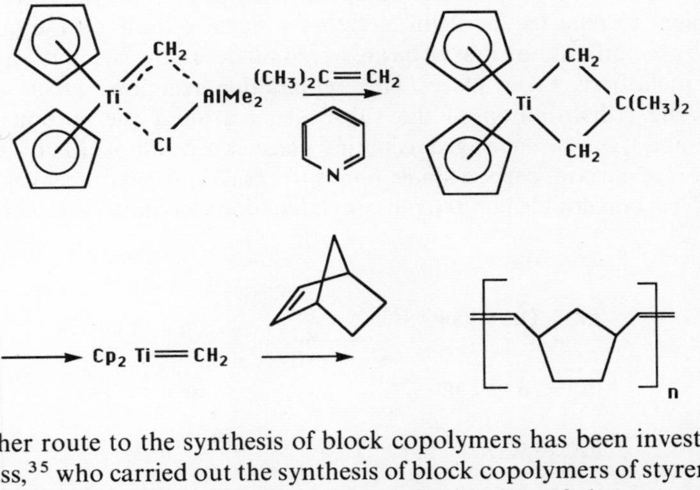

$$Cp_2Ti(\mu\text{-}CH_2)(\mu\text{-}Cl)AlMe_2 \xrightarrow[\text{pyridine}]{(CH_3)_2C=CH_2} Cp_2Ti(CH_2C(CH_3)_2CH_2)$$

$$\xrightarrow{\text{norbornene}} Cp_2Ti=CH_2 \longrightarrow [\text{polynorbornene}]_n \quad (3.5)$$

Another route to the synthesis of block copolymers has been investigated by Amass,[35] who carried out the synthesis of block copolymers of styrene and cyclopentene to elucidate the mechanism of initiation of the conventional metathesis catalyst. Using WCl_6 as catalyst and polystyryllithium as cocatalyst, the synthesis of block copolymers suggested that the metal alkyl was effectively the source of the metal carbene in the reaction and it was postulated that the active centre required metallation to be carried out by a metal alkyl that was essentially covalent and the propagating centre might have a bridge structure as shown in equation 3.6.

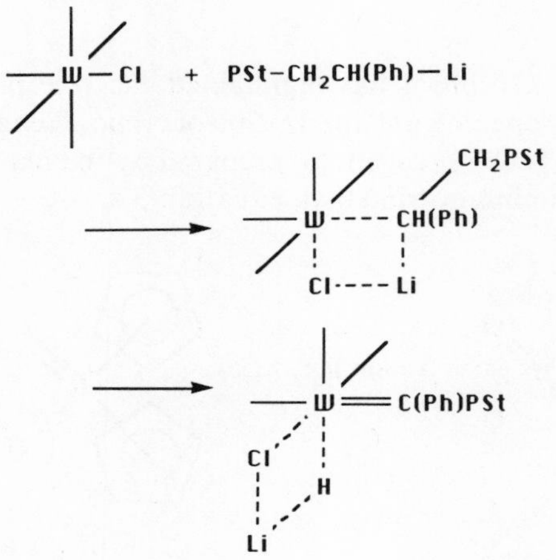

$$\begin{array}{c} \mathrm{W-Cl} \ + \ \mathrm{PSt-CH_2CH(Ph)-Li} \\ \downarrow \\ \mathrm{W----CH(Ph)} \\ \ \ \ \ \ \ \ |\ \ \ \ \ \ \ \ \ \ \ \ |\ \ \ \ \ \ \ \ \ \ \ \ \ \mathrm{CH_2PSt} \\ \mathrm{Cl----Li} \\ \downarrow \\ \mathrm{W=C(Ph)PSt} \\ \mathrm{Cl\ \ \ \ \ H} \\ \mathrm{Li} \end{array} \quad (3.6)$$

3.4 Mechanism of ring-opening metathesis polymerization

3.4.1 *Site of olefin metathesis*

To explain the mechanism of ring-opening metathesis polymerization it is convenient to refer to the olefin metathesis reaction itself. Of the questions posed by metathesis, the first to be answered relates to the site in the alkene at which metathesis takes place. This remarkable reaction, which involves simply the redistribution of the substituents around the carbon–carbon double bond, can be envisaged as taking place as a result of either a series of cleavages of carbon–carbon single bonds (transalkylation) or cleavage of the carbon–carbon double bonds (transalkylidenation), as shown in equations 3.7

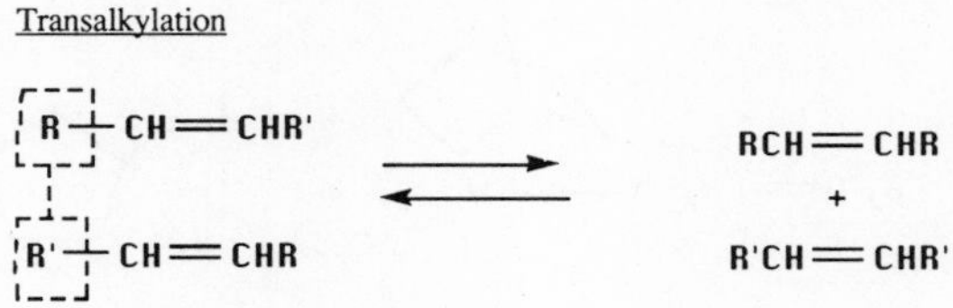

Transalkylation

$$\begin{array}{c} \mathrm{R{+}CH=CHR'} \\ \mathrm{R'{+}CH=CHR} \end{array} \rightleftharpoons \begin{array}{c} \mathrm{RCH=CHR} \\ + \\ \mathrm{R'CH=CHR'} \end{array} \quad (3.7)$$

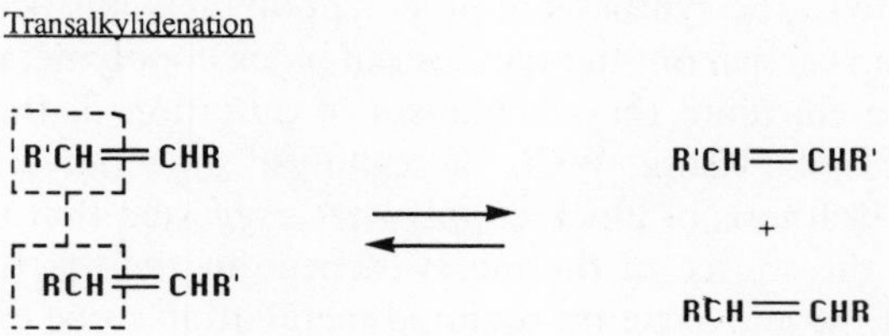

Transalkylidenation

$$\begin{array}{c} \mathrm{R'CH{+}CHR} \\ \mathrm{RCH{+}CHR'} \end{array} \rightleftharpoons \begin{array}{c} \mathrm{R'CH=CHR'} \\ + \\ \mathrm{RCH=CHR} \end{array} \quad (3.8)$$

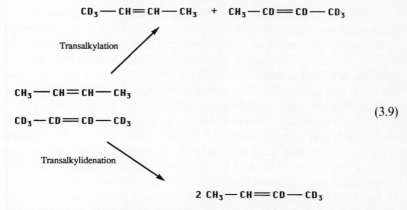

(3.9)

and 3.8. Calderon[36] devised a simple, but elegant, experiment to distinguish between these two pathways involving the metathesis of but-2-ene and perdeuteriated but-2-ene, as shown in equation 3.9. Analysis of the products of this metathesis reaction, which was catalysed by the $WCl_6/EtOH/AlEtCl_2$ complex, showed that only one new alkene was formed, 1,1,1,2-tetradeuteriobut-2-ene. It was therefore concluded that the metathesis reaction involved scission of the carbon–carbon double bond.

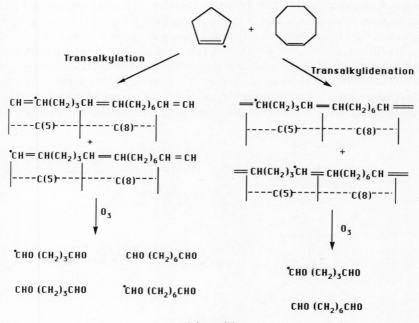

Scheme 3.1

Further evidence for the cleavage of the carbon–carbon double bond during the metathesis ring-opening polymerization of cycloalkenes came from a study of the copolymerization of cyclooctene and cyclopentene, the latter ^{14}C labelled at the carbon–carbon double bond.[37] During polymerization, the two possible products arising from either transalkylation or transalkylidenation are shown by Scheme 3.1. The copolymer obtained from the copolymerization of cyclooctene and cyclopentene was degraded by ozonolysis to the corresponding dials 1,8-octandial and 1,5-pentandial. Reduction of the dials to the diols was carried out with lithium aluminium hydride and these were separated and radiochemically assayed. An analysis of the C(5) and C(8) diols showed that all the radioactivity was associated with the 1,5-pentandiol. This observation was explainable only by the scission of the carbon–carbon double bond during metathesis; transalkylation would have resulted in an equal distribution of the radioactivity between the two diols. Taken together with the evidence provided by Calderon, the copolymerization studies conclusively prove that the olefin metathesis and ring-opening polymerization are mechanistically similar reactions occurring by a mechanism that involves scission of the carbon–carbon double bond.

3.4.2 Pairwise mechanisms

Only passing reference will be made here, for historical purposes, to the pairwise mechanisms that have been proposed for the metathesis reaction. The mechanism first proposed for metathesis was that of Calderon,[3] as shown in equation 3.10. It was proposed that the reaction proceeded by way of a (2 + 2) cycloaddition reaction between two alkene molecules to form a cyclobutane intermediate. Such reactions are symmetry forbidden according to the Woodward–Hofmann rules of conservation of orbital symmetry during chemical reactions. Sophisticated arguments were advanced in favour of the participation of the transition metal atom in the reaction that would convert a

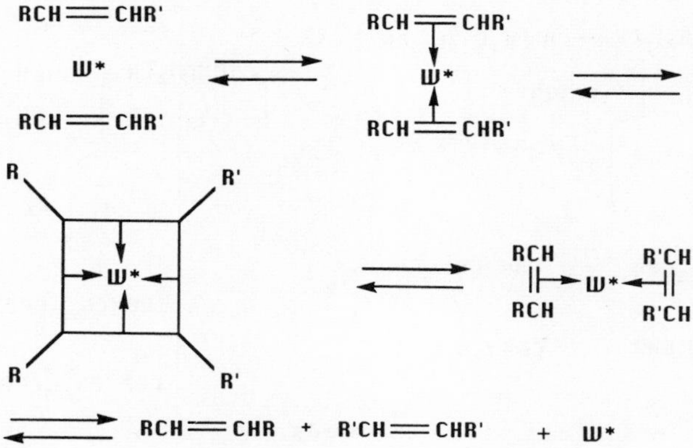

(3.10)

RING-OPENING METATHESIS POLYMERIZATION OF CYCLIC ALKENES

symmetry-forbidden into a symmetry-allowed reaction.[38] It was argued that the presence of the d-orbitals of the transition metal was a factor that permitted the reaction to proceed. A number of authors pointed out that cyclobutanes had never been isolated from metathesis reactions and it was therefore unlikely that they were to be considered to be real intermediates.[39]

Another speculative mechanism was that proposed by Grubbs,[40] which involved interaction of the transition metal and the two alkenes to form a metallocyclopentane ring which subsequently decomposed to form the new alkenes and a molecule of catalyst as shown in equation 3.11.

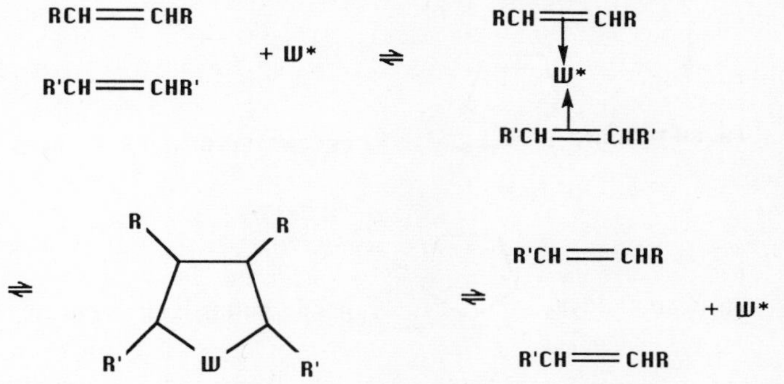

$$\text{(3.11)}$$

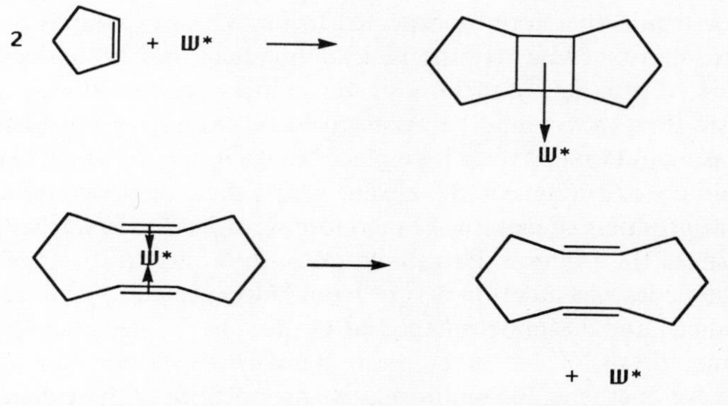

$$\text{(3.12)}$$

Each of these mechanisms need be modified only slightly to take account of a polymerization reaction, which is shown for the Calderon mechanism in equation 3.12, but it was information gained from the cross metathesis reactions between cyclic and linear alkenes that led to a reconsideration of the pairwise schemes as plausible explanations of the metathesis reaction.

3.4.3 Non-pairwise mechanisms

Herisson and Chauvin[41] were the first to challenge the concept of a pairwise mechanism for the metathesis reaction, although Calderon had reported that the polymerization of 1,5-cyclooctadiene produced not only polymer but also oligomers that were cyclic and formed a series $(C_4H_6)_n$, where $n \geqslant 3$. Herisson and Chauvin carried out an investigation of the products of the metathesis of cyclopentene and 2-pentene, there being three principal series of dienes, as shown in equation 3.13. Of these dienes, 2,7-decadiene homologues can be

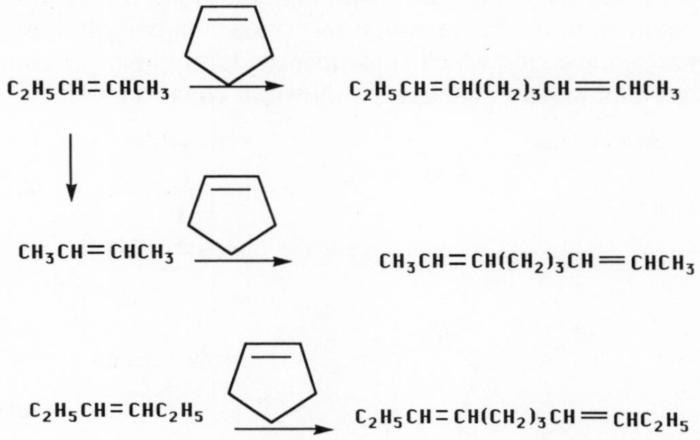

$$C_2H_5CH=CHCH_3 \longrightarrow C_2H_5CH=CH(CH_2)_3CH=CHCH_3$$

$$\downarrow$$

$$CH_3CH=CHCH_3 \longrightarrow CH_3CH=CH(CH_2)_3CH=CHCH_3$$

$$C_2H_5CH=CHC_2H_5 \longrightarrow C_2H_5CH=CH(CH_2)_3CH=CHC_2H_5 \qquad (3.13)$$

envisaged as arising as primary products from the cross-metathesis of cyclopentene and 2-pentene by a mechanism that could be pairwise. 2,7-Decadiene would therefore be expected to be one of the initial products of pairwise or non-pairwise metathesis. 2,7-Nonadiene and 3,8-undecadiene are also predicted products of metathesis, but if metathesis occurs by a pairwise mechanism then they would be expected to be secondary products because metathesis would first have to take place between two moles of 2-pentene to form a mixture of 2-butene and 3-hexene before these dienes could be formed. The initial products of metathesis therefore were seen to hold the key to the mechanism of the reaction. Extrapolation of the concentrations of the three important dienes as a function of time would be expected to go to zero for the 2,7-nonadiene and 3,8-undecadiene but not for the 2,7-decadiene, unless the rate of metathesis of 2-pentene were significantly faster than metathesis involving cyclopentene. In fact the relative proportions of these dienes did not vary significantly with time, contradicting the pairwise scheme.

These results led Herisson and Chauvin to predict a kinetic chain mechanism to explain the cross metathesis reaction. The propagating species was postulated to be a metallocarbene that also possessed a vacant co-ordination site at the transition metal atom. The first step in the reaction would be coordination of the carbon–carbon double bond at the transition metal atom, in a manner similar to the coordination that takes place during the propagation step in Ziegler–Natta polymerization. The coordinated monomer then reacts with the transition metal carbene to form a metallocyclobutane, which may revert to the previous form of the olefin coordinated to the metal carbene. If scission of the metallocyclobutane proceeds by route (a), as shown in equation 3.14, the new alkene and a new metal-carbene will form, whereas cleavage by route (b) will regenerate the original reactants.

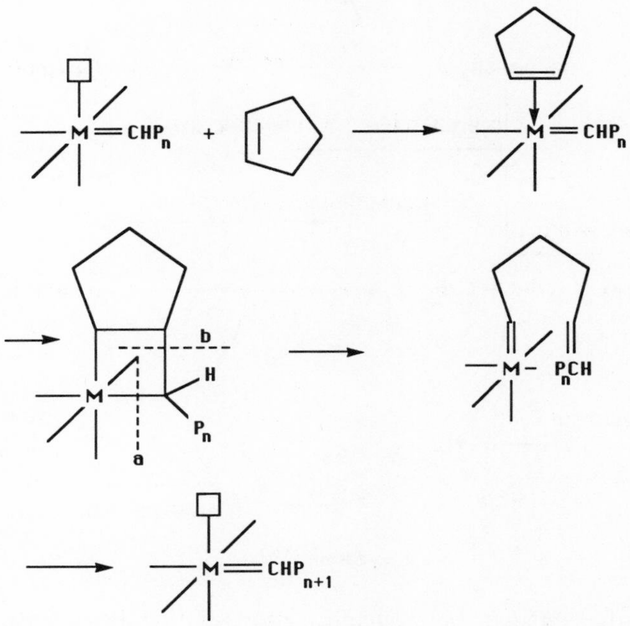

$$\tag{3.14}$$

Application of this kinetic chain mechanism to the reaction studied by Herisson and Chauvin demonstrated that all three dienes were equally likely to be formed in the initial stages of metathesis as shown in Scheme 3.2.

Katz and McGinnis[42] carried out a similar set of experiments that also concluded that the reaction could be a pairwise mechanism. Cyclooctene was cross metathesized with 2-butene and 4-octene. According to a pairwise mechanism the initial products could be only 2,10-dodecadiene and 4,12-hexadecadiene; the other product of primary metathesis by a pairwise scheme would be 2-hexene. The production of 2,10-tetradecadiene could only result as a secondary product by the cross metathesis of cyclooctene and 2-hexene. The important product ratios are then C_{14}/C_{12} and C_{14}/C_{16} which were determined as a function of reaction time. A plot was made of the product of these ratios against time. According to a pairwise mechanism the product of these ratios, $[C_{14}/C_{12}][C_{14}/C_{16}]$, should extrapolate against time to zero, whereas experimentally the extrapolated value was 4.1. This value can only be explained on the basis of a statistical incorporation of the alkene fragments as chain ends, unless the reaction between two moles of 2-hexene is so rapid that 2-butene and 4-hexene are present in the very early stages of the reaction.

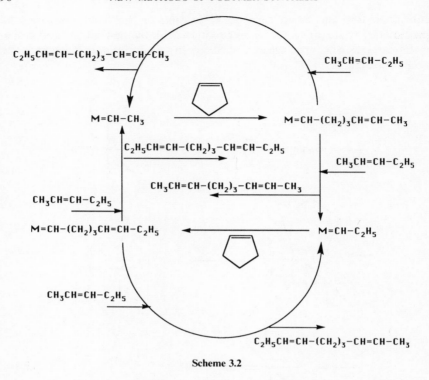

Scheme 3.2

Despite the fact that the mechanism proposed for the polymerization of cyclopentene invoked the presence of a metal carbene, no evidence was provided by Chauvin and Herisson, except that metal carbenes, such as the Fischer and Casey carbenes, may be prepared as stable compounds at low temperatures because the transition metal–carbon double bond is delocalized over the aromatic ring system. The reaction of an alkene with the Fischer carbene results in the elimination of carbon monoxide to create a vacant coordination site at which the alkene can coordinate. A Fischer carbene (**7**) may be modified by reaction with 3-butene-1-ol[43] to form (**8**).

$(CO_5)W = C(OCH_3) — C_6H_4\text{-p-Me} \rightarrow$
(**7**)
$(CO)_5W = C(O—(CH_2)_2CH = CH_2) — C_6H_4\text{-p-Me}$
(**8**)

This complex eliminates CO slowly to form the stable complex (**9**). Separation and characterization of (**9**) confirms the feasibility of the first step in the mechanism proposed by Herisson and Chauvin.

The original evidence in support of the second step in the reaction sequence, the formation of the metallocyclobutane, was mainly circumstantial because the synthesis of transition metal carbenes and metallocyclobutanes was rather

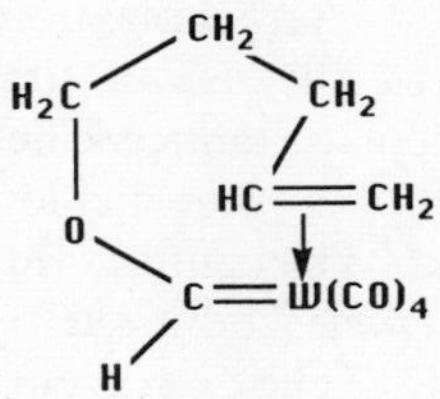

RING-OPENING METATHESIS POLYMERIZATION OF CYCLIC ALKENES 91

$$\begin{array}{c} \text{structure (9)} \end{array}$$

(9)

limited in scope. When the Casey carbene (**10**) was reacted with isobutene, the products of the reaction were explainable on the basis of the Chauvin mechanism, equation 3.15.[44]

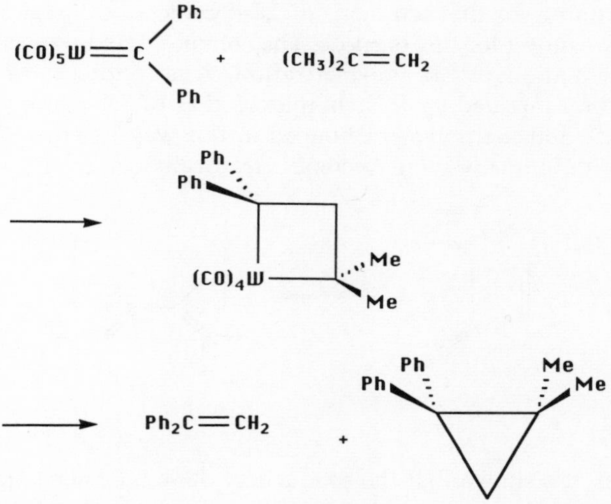

$$(CO)_5W=C(Ph)(Ph) + (CH_3)_2C=CH_2$$

$$\longrightarrow \text{metallocyclobutane intermediate}$$

$$\longrightarrow Ph_2C=CH_2 \quad + \quad \text{cyclopropane} \qquad (3.15)$$

The formation of 1,1-diphenylethylene was commensurate with the first metathesis product. Furthermore the formation also of 1,1-dimethyl-2-2-diphenylethylene was interpreted as evidence for the intermediate formation of the metallocyclobutane as an intermediate species in the reaction. In the reaction of neopentene with carbenic species, Schrock[45] showed that the active centre disappeared in a second-order manner and also generated products commensurate with the Chauvin mechanism. The destruction of the active centre in metathesis reactions may have an important equivalent polymerization termination reaction; Amass has shown that the termination of the polymerization of cyclopentene occurs in a second-order manner.[46]

3.4.4 Metallocyclobutanes

Following the success of the Chauvin mechanism in the interpretation of the polymerization of cyclopentene and the olefin metathesis reaction, the search began for the separation and characterization of the intermediates in the reaction. Grubbs has studied the use of titanium-based compounds for metathesis reactions, in particular titanocene derivatives. The reaction for the formation of complexes from titanocene dichloride and aluminium trimethyl was first described by Tebbe.[47] The complex (5) obtained in this reaction was essentially a titanium methylene stabilized by a complexing Lewis acid, aluminium dimethyl chloride. Tebbe's reagent was reacted under carefully controlled conditions with 2-butene[48] to form a stable titanocyclobutane. The *cis* isomer formed from 2-butene is less stable than the *trans* form; the structures of the metallocyclobutanes were confirmed by NMR spectroscopy.

1,1-Dicyclopentadienyl-*cis*-2,3-dimethyltitanocyclobutane, obtained when Tebbe's reagent is reacted with but-2-ene in the presence of pyridine, may be used as a catalyst for the metathesis of 2,8-decadiene, although it is a rather low-activity catalyst for this purpose. The complex may however be used for the rapid initiation of the polymerization of norbornene. An interesting polymerization initiated by this complex is that of 3,4-dipropylidenecyclobutene (11)[49] and the polymer obtained in this way by equation 3.16 may be doped with iodine, when it becomes electrically conductive.

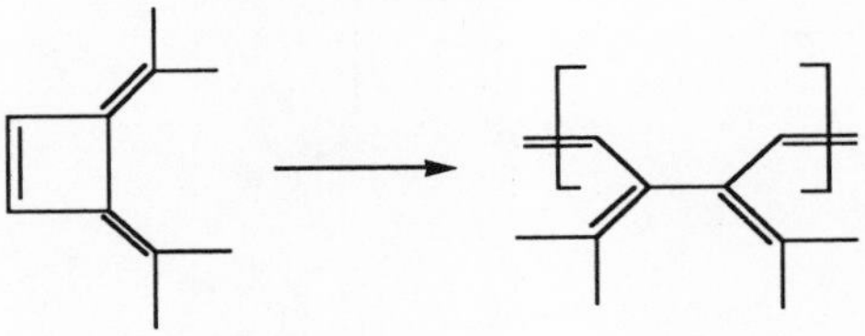

(3.16)

(11)

Recent work has shown that the products of these polymerization reactions are living polymers and that block copolymers may be synthesized by this technique. Examples of polymerization of this type are shown in Figure 3.4.

Remarkable though this route is for the synthesis of block copolymers, from a mechanistic point of view this work showed unequivocally that the metallocyclobutane is an important intermediate in the metathesis reaction. When a comparison is made between the titanium and tungsten derivatives it can be seen that the more stable form of the titanium derivative is the metallocyclobutane, whereas the more stable form of the tungsten derivative is the transition metal carbene; theoretical explanations for this behaviour have been advanced.

3.4.5 *Transition metal carbenes*

The other important intermediate proposed for the metathesis reaction is the transition metal carbene. Transition metal complexes involving a transition metal atom either with an alkyl group or with an alkylidene group are

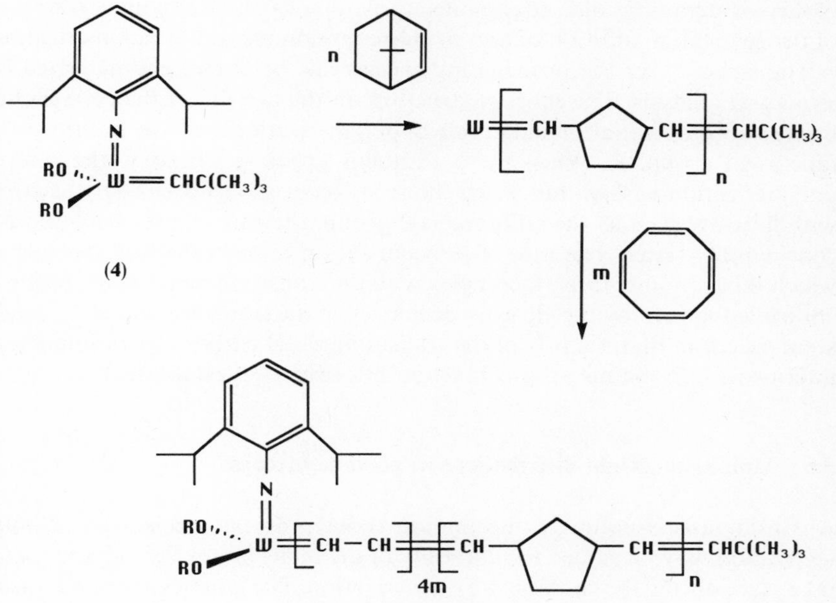

Figure 3.4

generally unstable unless some feature of the complex is designed to stabilize the molecule. The presence of some group that will delocalize the electrons of the carbene double bond or an electron donating ligand, such as a phosphine,[45,50] will improve the stability of such complexes. This explains the relative stability of the Fischer and Casey carbenes and hence the ability to synthesize these complexes. However, the very fact that these may be synthesized is because they tend to be stable and hence unreactive. Thus, although the initial products of metathesis can be explained on the basis of the Chauvin mechanism, efficient metathesis had not been achieved prior to the report by Schrock that a stable carbene (4) may be used to initiate the polymerization of norbornene derivatives. The polymerization is characterized by a fast rate of initiation and a stable propagating species that does not undergo termination.[12] The molecular weight of the polymer increases as a function of the fractional conversion of monomer to polymer, typical of a living polymerization. The polymerization is also characterized, as in all norbornene systems, by the absence of secondary metathesis reactions and hence the polydispersity of the polymer is narrow (< 1.1). Other monomers that may be polymerized include those monomers that have functional groups such as *endo-endo*-dicarbomethoxynorbornene,[51] provided that the monomer has a norbornene ring. The equivalent molybdenum initiator[52] may also be used to polymerize the bicyclic monomers (1) described by Feast for the synthesis of precursors to polyacetylene. The living character of these systems may also be exploited to synthesize block copolymers by way of the sequential addition of two or more monomers to the polymerization.

The structure of the propagating species has been well characterized by structural analysis. The effect of structure on the activity of the catalyst has been studied by variation of the rate of polymerization with the nature of the functional group, R. When the substituent group is t-butoxy, the rate of polymerization is slow, but when the substituent group is strongly electron withdrawing, such as the trifluoroacetyl group, the rate of polymerization is considerably faster. The rate of polymerization is controlled by the rate at which the monomer forms a complex with the transition metal atom, which is controlled in turn by the electron deficiency of the transition metal. It would seem therefore that the role of the transition metal carbene as an important intermediate in the metathesis reaction has been well-established.

3.5 Molecular weight distributions in polyalkenylenes

A considerable amount of information concerning the mechanism of polymerization may be gained from a study of the molecular weight changes that take place during the course of a polymerization. The kinetics and mechanism of polymerization will determine whether the number and weight average molecular weights increase or remain essentially constant during the course of conversion of the monomer to polymer.

Some of the earliest reports of the study of the molecular weight changes for the polymerization of cyclopentene with WCl_6/Ali-Bu_3 showed that the polymer produced in the early stages of the reaction was of very high molecular weight and, as shown earlier by Chauvin, this could be explained by a kinetic chain mechanism. However, a more surprising result was that the number average molecular weight decreased with conversion of the monomer.[53,54] Gel permeation chromatographic analysis of the polymers showed that the polymers had bimodal molecular weight distributions. The decrease in number average molecular weight was associated with an increase in the ratio of the weight of the low molecular to high molecular weight material. It was shown by Hoecker and Calderon that the low molecular weight material is cyclic and that when the monomer is 1,5-cyclooctadiene, the oligomers are cyclics of multiples of the C4 units.[26,55]

Two explanations have been proposed for the production of the distribution, one being kinetic[56] and the other thermodynamic.[55] The low molecular weight material was produced in the early as well as the later stages of the polymerization and, since the reactions between WCl_6 and cyclopentene lead to a mixture of oxidation states of the tungsten atom,[57] it is possible that during the initial stages of the reaction the bimodal distribution is formed as a result of polymerization by two kinetically distinct species. The initial weight ratio of low to high molecular weight material increased as the time of pre-

RING-OPENING METATHESIS POLYMERIZATION OF CYCLIC ALKENES 95

reaction of the catalyst with the monomer was increased. The increase in reaction time was believed to increase the proportion of the catalyst that existed in the low molecular weight producing form. However, in all polymerizations the molecular weight distribution eventually tends to the molecular weight distribution associated with thermodynamic control.

A special feature of the metathesis polymerization of cyclopentene is the fact that the double bond formed in the polymer is also susceptible to secondary metathesis reactions, unlike the double bond formed when norbornene polymerizes. Secondary metathesis reactions may be either intermolecular or intramolecular in nature. If the secondary metathesis reaction is as described in Scheme 3.3, i.e. an intramolecular reaction, the overall number average molecular weight is unaffected, and only the molecular weight distribution is affected by broadening. The process of secondary metathesis by intramolecular reactions is a randomizing process that may generate a statistical distribution from a narrow distribution polyalkenylene. When the secondary metathesis reaction is intramolecular in nature the reaction will lead to the production of a lower molecular weight polymer and a cyclic oligomer.

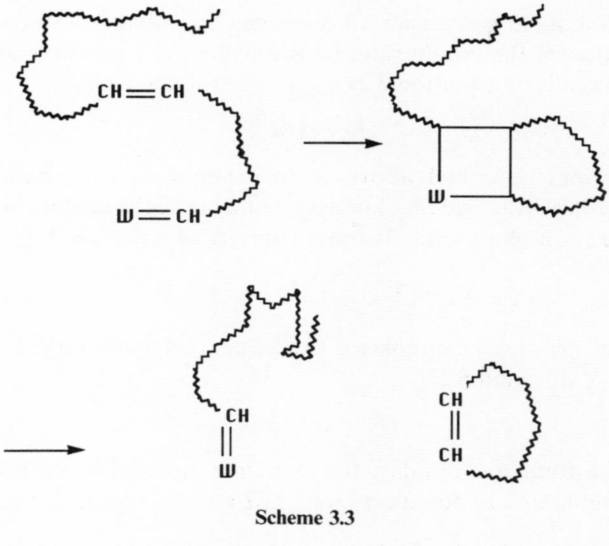

Scheme 3.3

If ring-opening polymerization of a monomer such as cyclooctene is carried out at low concentrations of monomer, the thermodynamic control favours the formation of the oligomers. Mass spectroscopic studies, particularly on the hydrogenated products, have shown that a homologous series of cyclic oligomers is formed. Above the equilibrium concentration of monomer the relative concentrations of the oligomers are independent of the initial concentrations of the monomer and catalyst; only a slight dependence of the ratio on temperature is found and as the initial concentration of monomer is increased the ratio of polymer to oligomers also increases.

The equilibrium between open-chain polymer and each cyclic oligomer may be represented by equation 3.17:

$$A—M_y—B \quad A—M_{y-x}—B + M_x \tag{3.17}$$

The equilibrium constant (K_x) for the formation of the cyclic oligomer will be given by equation 3.18:

$$K_x = [M_x][A—M_{y-x}—B]/[A—M_y—B] \tag{3.18}$$

If the probability that any carbon–carbon double bond, selected at random, has been polymerized is given by p, then the ratio of the concentrations of (y-x)-mer to y-mer is given by equation 3.19:

$$[A—M_{y-x}—B]/[A—M_y—B] = p^{y-x}/p^y = 1/p^x \tag{3.19}$$

Therefore the equilibrium constant (K_x) is given by equation 3.20.

$$K_x = [M_x]/p^x \tag{3.20}$$

When the fractional conversion of monomer to polymer approaches unity, then the value of the equilibrium constant for the formation of the cyclic oligomer is given by equation 3.21:

$$K_x = [M_x] \tag{3.21}$$

The equilibrium described above is an example of ring-chain equilibria described by Jacobson and Stockmayer,[58] and hence the equilibrium constant for the formation of a cyclic oligomer is given by equation 3.22:

$$K_x = [3/2]^{3/2} 1/xL.[1/\langle r^2 \rangle]^{3/2} \tag{3.22}$$

The value of $\langle r^2 \rangle$ may be replaced by the characteristic ratio for the cyclic oligomer (c_x), equation 3.23:

$$c_x = \langle r^2 \rangle / vxl^2 \tag{3.23}$$

where v is the number of bonds in the monomer unit and l is the average bond length. Combination of equations 3.21, 3.22 and 3.23 leads to equation 3.24.

$$[M_x] = [3/2]^{3/2} 1/L.[1/vl^2 c_x^{3/2}].x^{-5/2} \tag{3.24}$$

The equilibrium concentrations of the cyclic oligomers during the polymerization of cycloalkenes has been determined by gel permeation chromatography, and plots of $\ln[M_x]$ against x show a slope of $-5/2$ in agreement with the Jacobson–Stockmayer theory on ring-chain equilibria. The intercepts obtained show the expected dependences on the characteristic ratio.

3.6 Stereochemistry of ring-opening metathesis polymerization

One of the remarkable features of ring-opening metathesis polymerization reactions is that the carbon–carbon double bond of the monomer is retained during the course of the polymerization reaction. The carbon–carbon double bond is therefore one source of stereoisomerism in the polymer and may be either *cis* or *trans* whether the polymer is obtained from a monocyclic or a bicyclic monomer. When the polymer is obtained by the polymerization of a bicyclic monomer the situation is further complicated by the fact that the orientation of the in-chain carbocyclic ring is another source of stereoisomerism. The polymerization of norbornene is an example of the latter type, as shown in equation 3.25.

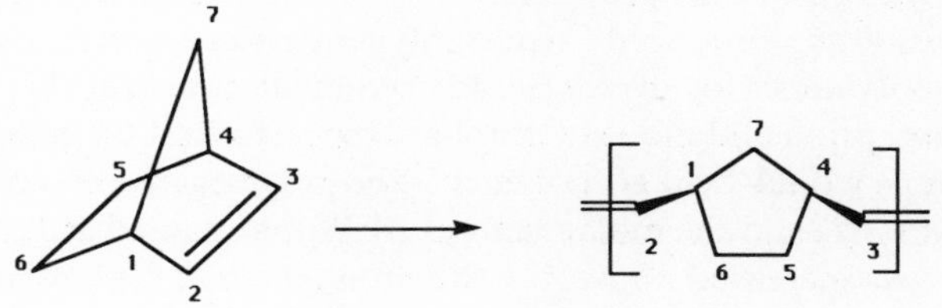

(3.25)

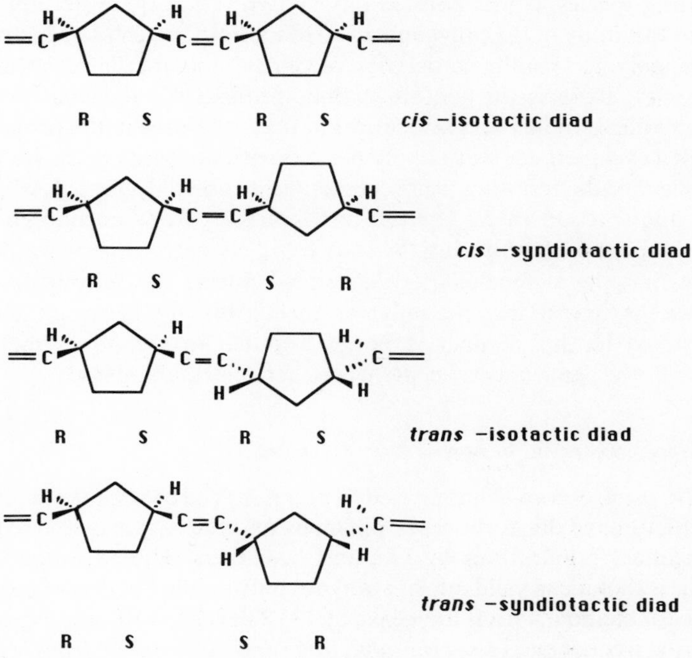

Figure 3.5

During the course of the polymerization the carbon atoms 2 and 3 become part of in-chain double bonds, *cis* or *trans*, and 1 and 4 are chiral centres along the polymer chain, whose absolute configurations may be either **R** or **S**. It can readily be seen that the absolute configurations of carbon atoms, 1 and 4, must be opposite for each cyclopentane ring. Various combinations of chiral centres are seen to be possible depending upon the orientation of the cyclopentane ring with respect to the elongated chain and the isomerism of the carbon–carbon double bond. If the discussion is restricted to diad combinations, these are shown for all-*cis* and all-*trans* polymers in Figure 3.5.

3.6.1 *Stereoisomerism of polyalkenylenes*

The initial studies of the polymerization of cyclopentene suggested that tungsten catalysts produced *trans*-polypentenylene and molybdenum catalysts *cis* polymers. However, a broad classification such as this is bound to be open to exception and indeed Pampus[20] reported that the polymerization of cyclopentene would yield *cis* polymer if the polymerization were catalysed by $WF_6/Al_2Et_3Cl_3$ and the molar ratio of Al:W maintained at 1:1. When the molar ratio was increased above 1:1 the product was of increasingly higher *trans* content. No attempt was made to explain this effect except to say that the stereochemistry of the polymer was determined by the structure of the propagating species. It was pointed out by Ivin[58] that there were problems related to the study of the polymerization of monomers such as cyclopentene since the polymer readily undergoes secondary metathesis reactions and, consequently, there is no guarantee that the final stereochemistry of the carbon–carbon double bond is identical to that generated in the propagation step. When cyclopentene was copolymerized with norbornene the fraction of *cis* double bonds resulting from pentenylene units decreased with time whereas the fraction of *cis* double bonds arising from norbornene units remained fairly constant during the course of polymerization. It would seem therefore that the stereochemistry generated during the propagation with norbornene is 'frozen' into the polymer during this step because the double bond derived in this manner is not susceptible to secondary metathesis reactions in the same way as pentenylene derived double bonds.

3.6.2 *Stereoisomerism in norbornene polymers*

^{13}C NMR spectroscopy[24] has proved to be a powerful tool for the elucidation of the structures of the norbornene polymers and the reader is referred to the many excellent publications by Ivin and coworkers. The advantage of this technique is that it can yield information not only on the fraction of *cis* (r_c) and *trans* (r_t) double bonds from the peaks at 133.88 and 133.10 ppm respectively, but also the frequency of *cis–cis*, *trans–trans* and *cis–trans* or *trans–cis* diads because the chemical shift of the chiral carbons (1 and 4) depends not only on the stereostructure of the nearest double bond but on the structure of the next nearest also. This technique therefore gives an indication of the frequency distribution of structural diads and hence any tendency for the formation of random or blocky polymers.

The polynorbornylene may be considered to be a copolymer of *cis* and *trans* monomer units and it is possible to define the copolymerization reactivity ratios $r_c = (2)cc/(2)ct$ and $r_t = (2)tt/(2)tc$ where $(2)cc$ represents the carbon atom (2) that is part of a *cis–cis* diad etc. A tendency to blockiness occurs when $r_c \cdot r_t > 1$ and a statistical distribution occurs when $r_c \cdot r_t = 1$. Ivin *et al.* have studied the synthesis of polynorbornylene with a range of transition metal catalysts and have shown that the catalysts that produce low *cis* content polymers have a tendency to produce a statistical distribution of double bonds whereas as the *cis* content increased the blockiness increased also. Thus there is a tendency for *cis* units to occur in blocks and in the main fast-acting catalysts produce low *cis* content polymers and slow-acting systems produce high *cis* contents.

The studies have shown that the microstructure of the polymer depends not only on the transition metal atom but also on the ligands surrounding that atom. To account for the variation of the stereochemistry of the polymer, two propagating species were postulated and that these were in equilibrium with one another although when the catalyst is a high *cis*-forming catalyst the rate of interchange between these two forms is very slow.[59] Thus once a catalyst enters a *cis*-producing mode it continues to propagate in that mode until disturbed from that mode. The structures proposed are shown as **(12)** and **(13)**.

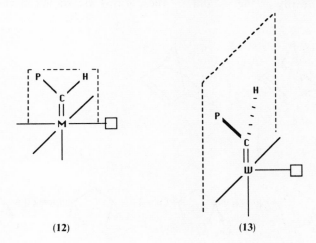

(12)   (13)

3.6.3 Substituted norbornenes

Although the polymerization of norbornene yields a polymer in which the microstructure generated during polymerization does not change significantly afterwards, the orientation of the cyclopentane ring, and hence the tacticity of the polymer, cannot be discerned from the NMR spectrum of polynorbornene itself. To study the effects of catalyst structure on the ring tacticity, another monomer, in which *m*- and *r*- diads are distinguishable from one another, has to be used. A suitable monomer for these studies was found to be 1-methyl-bicyclo[2.2.1.]hept-2-ene,[60] as shown in Figure 3.6. During polymerization, in addition to the isomerism associated with the double bond, there is the stereoisomerism associated with the orientation of the ring and the possibility of the formation of tail-to-head, head-to-head and tail-to-tail structures. The ^{13}C NMR spectrum of this polymer is now very complex as the alkenic carbon chemical shift depends not only on the nature of the double bond (*cis* or *trans*) but also on the arrangements of the two repeat units with which it is associated, HH, TH (HT) or TT. Experiments have shown that all arrangements are found except *cis*-HH.

It is now necessary to define the bias (B) of the polymer as:

$$B = \frac{HT + TH}{HH + TT}$$

The bias is a measure of the tendency for the polymerization to propagate in a head-to-tail, or a normal, manner. In polymerizations of this monomer, the propagating carbene may have the methyl group adjacent to the carbene (P_H) or removed from the carbene (P_T). If the metal centre is assumed to be chiral, not an unreasonable assumption considering its structure, then each of the propagating metal carbenes can exist in one of two enantiomeric forms assigned arbitrarily as the + and −; these forms are shown as (14) to (17).

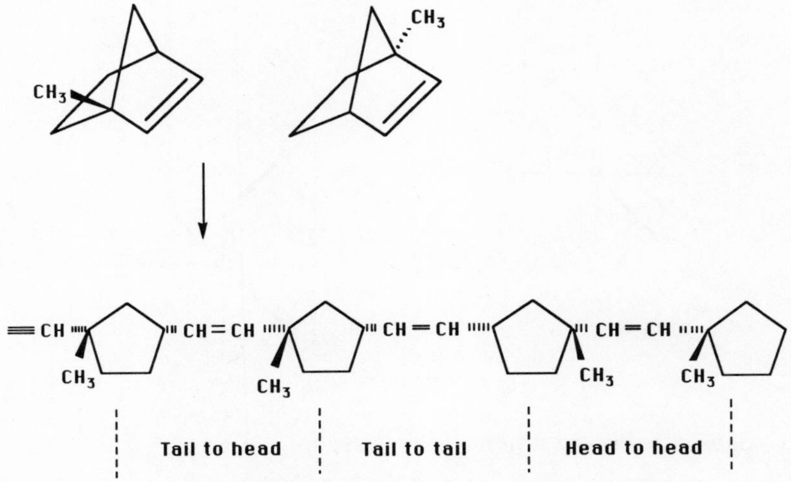

Figure 3.6

Inspection of the monomer also shows that the monomer is chiral and therefore exists in two enantiomeric forms which are also shown in Figure 3.6.

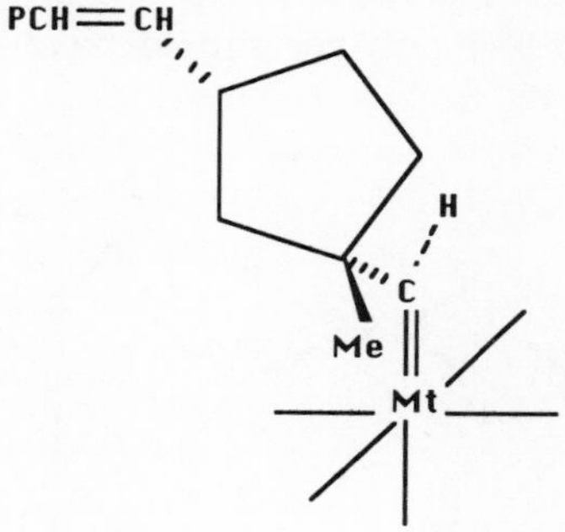

(14) P_H^+

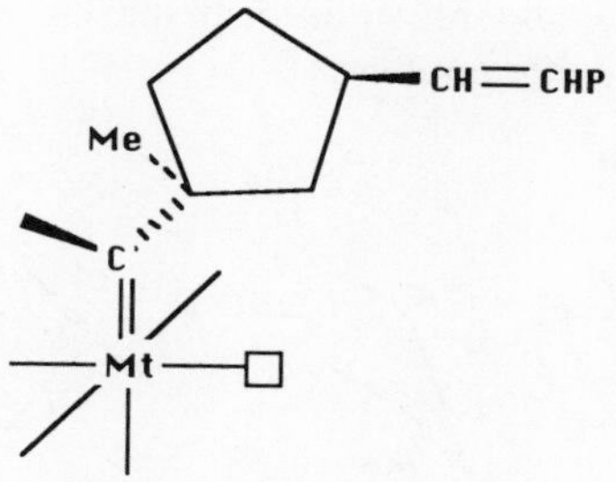

(15) P_H^-

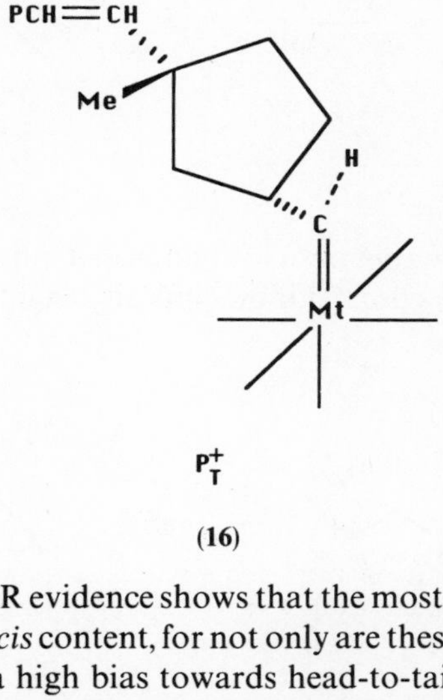

(16) P_T^+

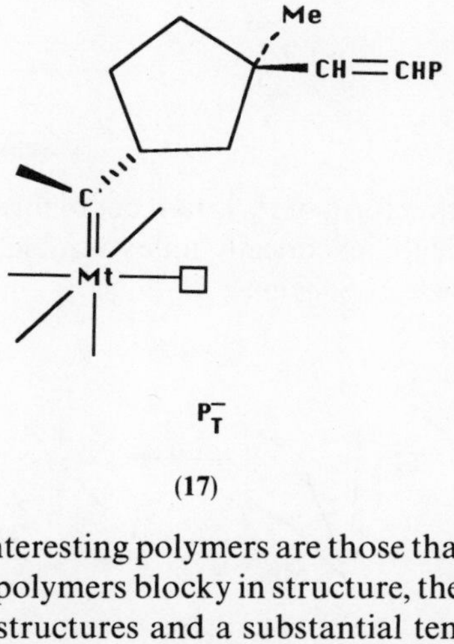

(17) P_T^-

NMR evidence shows that the most interesting polymers are those that have a high *cis* content, for not only are these polymers blocky in structure, they also show a high bias towards head-to-tail structures and a substantial tendency towards syndiotactic placements of the cyclopentane rings. If *trans* placements occur they are associated with head-to-head diads and when they occur they are usually preceded by a head-to-tail *trans* reaction and followed by a tail-to-tail placement, which can be either *cis* or *trans* forming.

In reactions between two substrates, both of which can exist in two enantiomeric forms, the reactants of similar chirality react with one another; thus (+) enantiomers react together as do (−) forms. During polymerization to form the *cis* polymer with regular head-to-tail placements, the propagating species alternates between the two enantiomeric forms. It seems reasonable therefore to suggest that the racemic mixture of MNBE reacts in a similar manner, i.e. that there is an alternating incorporation of the (+) and (−) enatiomeric forms of the monomer. This mechanism of propagation also predicts, as shown in Scheme 3.4, that the polymer should have a high syndiotactic content.

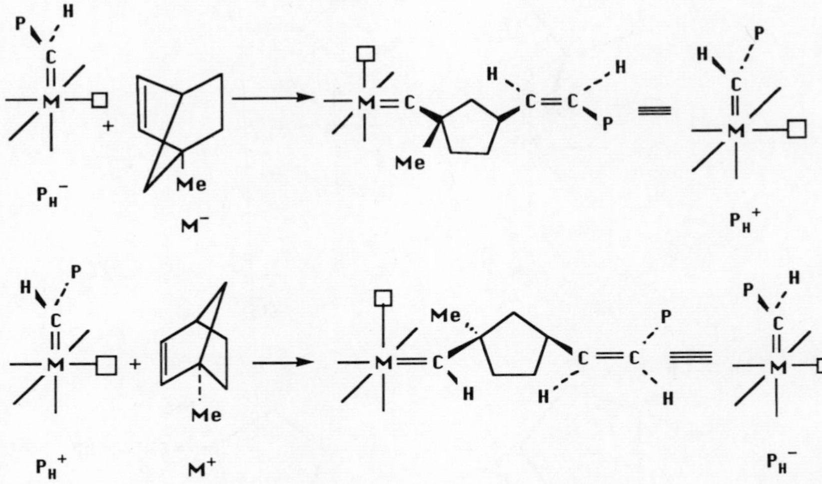

Scheme 3.4

If either form of P_H is to react to form a *trans* double bond then it must do so in the stereochemically unfavourable reaction with the opposite enantiomer, as shown in Scheme 3.5.

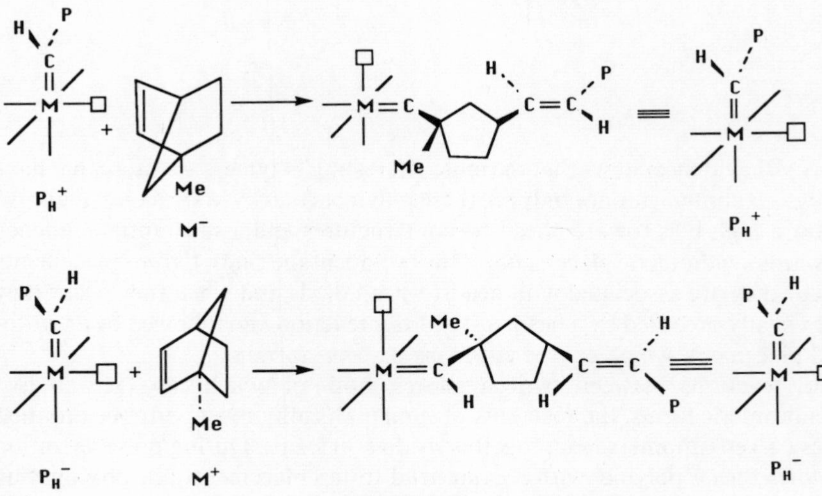

Scheme 3.5

RING-OPENING METATHESIS POLYMERIZATION OF CYCLIC ALKENES 103

The formation of a *trans* double bond is more likely therefore to take place by way of one of the reactions that leads to a head-to-head or tail-to-tail placement, as shown in Scheme 3.6. The success that the study of the stereochemistry of norbornene has had in predicting the course of the polymerization has undoubtedly shed considerable light on the carbene mechanism of metathesis polymerization.[61]

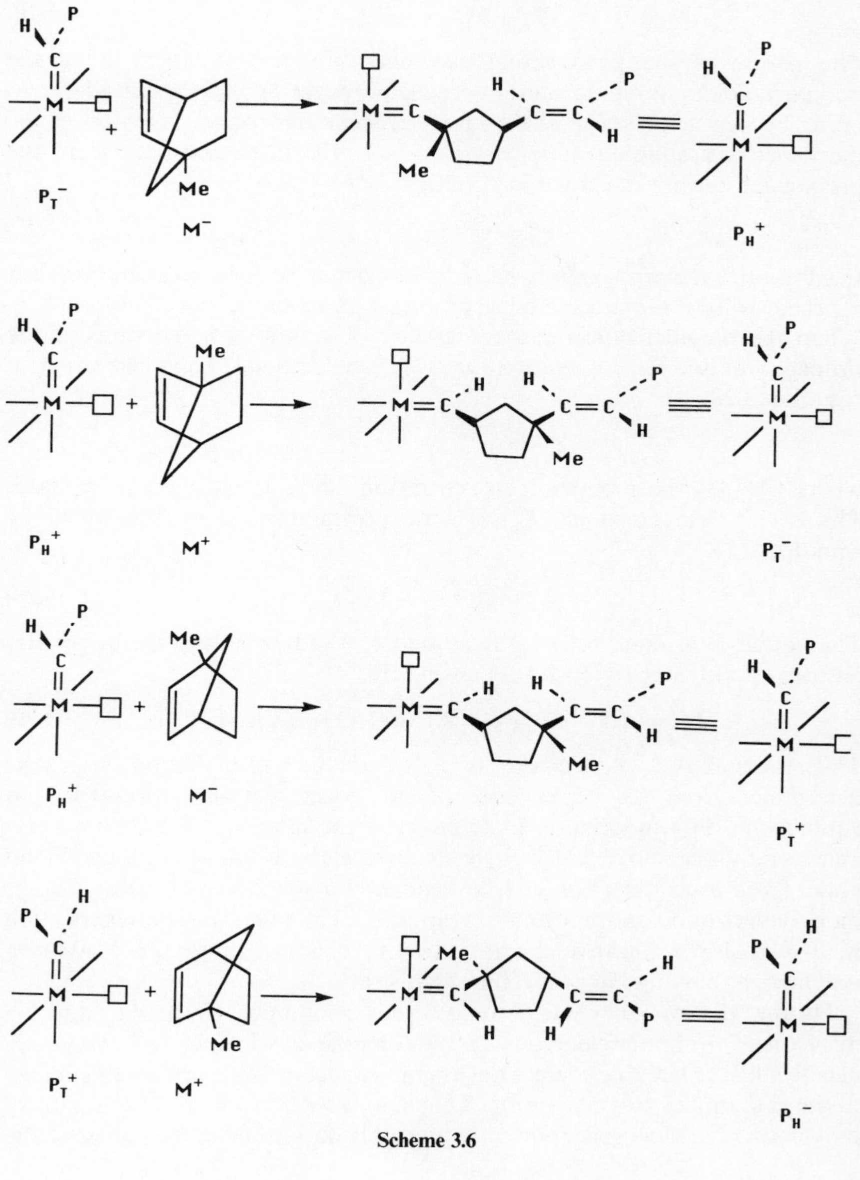

Scheme 3.6

3.7 Thermodynamics of ring-opening polymerization

The laws of thermodynamics, describing only the feasibility of chemical changes, may be applied to polymerization of any monomer and the polymerization of cyclic systems in particular. The propagation reaction may be summarized for any system by equation 3.26.

$$P_n^* + M \longrightarrow P_{n+1}^* \tag{3.26}$$

The nature of the propagating end plays no part in determining the polymerizability of the monomer because propagation may be considered to occur by way of cleavage of the chain at some mid point, insertion of the monomer and subsequent recombination of the new species to form the propagated centre, as shown in equation 3.27.

$$P_n^* = P_{p+q}^* \longrightarrow P_p \sim + \sim P_q^* \longrightarrow P_{p+1} \sim + \sim P_q \longrightarrow P_{p+1+q}^* = P_{n+1}^* \tag{3.27}$$

In addition to the propagation reaction, which may be assigned a rate constant k_p, there will be the associated depropagation reaction, rate constant k_{-p}. When the polymerization reaches a state of equilibrium the rates of the propagation and depropagation reactions are identical. Then according to equation 3.28

$$k_p[M]_e[P_n^*] = k_{-p}[P_{n+1}^*] \tag{3.28}$$

where $[M]_e$ is the monomer concentration when equilibrium is attained. The equilibrium constant (K_p) for the polymerization is then given by equation 3.29.

$$K_p = k_p/k_{-p} = 1/[M]_e \tag{3.29}$$

The equilibrium constant (K_p) is related to the free energy, enthalpy and entropy of polymerization by equation 3.30.

$$\Delta G = \Delta H - T\Delta S = - RT \ln (K_p) = RT \ln ([M]_e) \tag{3.30}$$

Thus the enthalpy and entropy of polymerization of a monomer may be determined from the dependence of the concentration of monomer at equilibrium on temperature. In the case of the majority of addition polymerizations there is a change in hybridization of the atoms of the double bond undergoing polymerization and consequently a significant enthalpy change on polymerization, that is usually exothermic. The entropy of polymerization is also usually a negative quantity and so it is not unusual for addition polymerizations to exhibit a ceiling temperature.

During the polymerization of a cyclic monomer by a ring-opening mechanism, the bonds cleaved in the monomer and formed in the polymer are usually identical, so that there is not a significant contribution from this source to the enthalpy of polymerization. The major contribution to the enthalpy of polymerization of a cyclic monomer is usually derived from the release of the ring-strain energy that takes place during the course of polymerization. Thus Dainton and Ivin[62] have shown that the ring strain energy of the monomer is the most important thermodynamic factor to be considered when assessing the polymerizability of a cyclic monomer. Those monomers that display no ring strain energy, such as cyclohexene, are not usually to be polymerized by a ring-opening process, whereas the monomer will exhibit a reasonably high ceiling temperature, even if the ring strain energy is only small. Thus cyclopentene will polymerize by a ring-opening metathesis reaction, the equilibrium concentration of monomer being of the order of 1 molar at 25 °C. Of the other monocyclic monomers that undergo metathesis ring-opening polymerization, cyclooctene and cyclooctadiene are such strained rings that the equilibrium concentration of monomer is extremely low. These features are displayed by other groups of monomers that polymerize by ring-opening mechanisms and in some cases the formation of cyclic compounds may take place from open-chain molecules, e.g. 2-hydroxycarboxylic acids might be expected to form six-membered rings rather than polymerize. Ivin has published the ring strain energies of a series of monocyclic alkanes which acts as a useful guide to the polymerizability of cyclic monomers. The inability to polymerize six-membered rings, which are able to adopt strain-free conformations, does not necessarily apply to the polymerization of bicyclic monomers that contain the reactive functional group in a five- or six-membered ring. Again the ring-opening metathesis polymerization of cycloalkenes provides a suitable example of this effect since norbornene, 2,2,1-bicyclohept-2-ene, may be polymerized to almost 100% conversion using a broad range of metathesis catalysts; even though the monomer contains only five- and six-membered rings the bridgehead carbon atom constrains the ring to such an extent that the strain in the molecule is quite considerable.

References

1. Ivin, K.J. *Olefin Metathesis*, Academic Press, London (1983); Dragutan, V., Balaban, A.T. and Dimonie, M. *Olefin Metathesis and Ring Opening Polymerization of Cyclo-olefins* (2nd. edn), Wiley-Interscience, New York (1985); Rooney, J.J. and Steward, A. *Chem. Soc. Specialist Periodical Reports, Catalysis* **1**, 277 (1977).
2. Banks, R.L., and Bailey, G.C. *Ind. Eng. Chem. Prod. Res. Dev.* **3**, 170 (1964).
3. Calderon, N., Chen, N.Y. and Scott, K.W., *Tetrahedron Letters* **34**, 3327 (1967).
4. Edwards, J.H. and Feast, W.J. *Polymer* **21**, 595 (1980).
5. Patton, P.A. and McCarthy, T.J. *Macromolecules* **20**, 778 (1987).
6. Feast, W.J. and Wilson, B. *J. Molec. Catal.* **8**, 277 (1980).
7. Pampus, G., Witte, J. and Hoffmann, M. *Rev. Gen. Caout. Plast.* **47**, 1343 (1970).
8. Reif, L. and Hoecker, H. *Makromol. Chem., Rapid Commun.* **2**, 745 (1981).
9. Boelhouwer, C. and Mol, C.J. *Prog. Lipid Res.* **24**, 243 (1985).
10. Amass, A.J., Dale, A.L. and Tighe, B.J. *Makromol. Chem.* **189**, 515 (1988).
11. Novak, B.M. and Grubbs, R.H. *J. Amer. Chem. Soc.*, **110**, 960 (1988).
12. Schrock, R.R., Grubbs, R.H., Feldman, J. and Cannizzo, L.F. *Macromolecules* **20**, 1169 (1987).
13. Eleuterio H.S. *US Patent* (1957) 3 074 918; *Chem. Abstr.* **55**, 16005 (1961).
14. Verkuijlen, E., Kapteijn, F., Mol, J.C. and Boelhouwer, C. *J. Chem. Soc., Chem. Commun.* 198 (1977).

15. Amass, A.J. and McGourtey, T.A. *Eur. Polym. J.* **16**, 235 (1980).
16. Schrock, R.R., Krouse, S.A., Knoll, K., Feldman, J., Murdzek, J.S. and Yang, D.C. *J. Mol. Cat.* **46**, 243 (1988).
17. Wallace, K.R., Liu, A.H., Dewan J.C. and Schrock, R.R. *J. Amer. Chem. Soc.* **110**, 4964 (1988).
18. Gilliom, L.R. and Grubbs, R.H., *J. Amer. Chem. Soc.* **108**, 733 (1986).
19. Farona, M.F., Lofgren, P.A. and Woon, P.S. *J. Chem. Soc., Chem. Comm.* 246 (1974).
20. Guenther, P., Haas, F., Marwede, G., Nuetzel, K., Oberkirch, W., Pampus, G., Schoen, N. and Witte, J. *Angew. Makromol. Chem.* **14**, 87 (1970).
21. Wallace, K.R. and Schrock, R.R. *Macromolecules* **20**, 448 (1987).
22. Natta, G., Dall'Asta, G. and Motroni, G. *J. Polym Sci. Polym. Lett.* **B2**, 349 (1964).
23. Michelotti, F.W. and Keaveney, W.P. *J. Polym Sci.* **A3**, 895 (1965).
24. Ivin, K.J., Laverty, D.T., O'Donnell, J.H., Rooney, J.J. and Stewart, C.D. *Makromol. Chem.* **180**, 1989 (1979).
25. Natta, G., Dall'Asta, G., Mazzanti, G. and Motroni, G. *Makromol. Chem.* **69**, 163 (1963).
26. Calderon, N., Ofstead, E.A. and Judy, W.A. *J. Polym Sci.* A(1), **5**, 2209 (1967).
27. Dall'Asta, G. and Montroni, G. *Chem. Abstr.* **66**, 19005 (1967).
28. Katz, T.J. and Acton, N. *Tetrahedron Letters* 4251 (1976).
29. Katz, T.J., Lee, S.J. and Acton, N. *Tetrahedron Letters* 4247 (1976).
30. Katz, T.J., Lee S.J. and Shippey, M.A. *J. Mol. Catal.* **8**, 219 (1980).
31. Kress, J., Osborne, J.A., Green, R.M.E., Ivin, K.J., Rooney, J.J. *J. Chem. Soc., Chem. Commun.* 874 (1985).
32. Kress, J. and Osborne, J.A. *J. Amer. Chem. Soc.* **109**, 3953 (1987).
33. Kress, J., Osborne, J.A., Green, R.M.E., Ivin, K.J. and Rooney, J.J. *J. Amer. Chem. Soc.* **109**, 899 (1987).
34. Kress, J., Osborne, J.A., Amir-Ebrahimi, V., Ivin, K.J. and Rooney, J.J. *J. Chem. Soc., Chem. Comm.* 1164 (1988).
35. Amass, A.J. and Gregory D. *Brit. Polym J.* **19**, 263 (1987).
36. Calderon, N., Ofstead, E.A., Ward, J.P., Judy, W.A. and Scott, K.W. *J. Amer. Chem. Soc.* **90**, 4133 (1968).
37. Dall'Asta, G. and Montroni, G. *Eur. Polym. J.* **7**, 707 (1971).
38. Mango, F.D. and Schachtschneider, J.H. *J. Amer. Chem. Soc.* **93**, 1123 (1971).
39. Lewandos, G.S. and Pettit, R. *Tetrahedron Letters* 789 (1971).
40. Biefield, C.G., Eick, H.A. and Grubbs, R.H. *Inorg. Chem.* **12**, 2166 (1973).
41. Herisson, J.L. and Chauvin, Y. *Makromol. Chem.* **141**, 161 (1970).
42. Katz, T.J. and McGinnis J. *J. Amer. Chem. Soc.* **97**, 1592 (1975).
43. Casey, C.P. and Shusterman, A.J. *J. Molec. Catal.* **8**, 1 (1980).
44. Casey, C.P., Tuinstra, H.E. and Saeman, M.C., *J. Amer. Chem. Soc.* **98**, 608 (1976).
45. Schrock, R.R., Rocklage, S., Wengrovius, J., Rupprecht, G. and Fellman, J. *J. Molec. Catal.* **8**, 73 (1980).
46. Amass, A.J. and Zurimendi, J.A. *Eur. Polym. J.* **17**, 1 (1981).
47. Tebbe, F.N., Parshall, G.W. and Reddy, G. *J. Amer. Chem. Soc.* **100**, 3611 (1978).
48. Howard, T.R., Lee, J.B. and Grubbs, R.H. *J. Amer. Chem. Soc.* **102**, 6876 (1980).
49. Ott, K.C., Lee, J.B. and Grubbs, R.H. *J. Amer. Chem. Soc.* **104**, 2942 (1982).
50. Wengrovius, J.H., Schrock, R.R., Churchill, M.R., Missert, J.R. and Youngs, W.J. *J. Amer. Chem. Soc.* **102**, 4515 (1980).
51. Murdzek J.S. and Schrock, R.R. *Macromolecules* **20** (1987).
52. Knoll, K., Krouse, S.A. and Schrock, R.R. *J. Amer. Chem. Soc.* **110**, 4424 (1988).
53. Pampus, G., Witte J. and Hoffmann, M. *Rev. Gen. Caout. Plast.* **47**, 1343 (1970).
54. Amass, A.J. and McGourtey, T.A. *Proceedings of 'Catalyse par Coordination (Villeurbanne, Lyon, France)'* p. 181 (1974).
55. Hoecker, H., Reiman, W., Reif L. and Riebel K. *Rec. Trav. Chim. Pays-Bas* **96**, M47 (1977).
56. Amass, A.J. and Zurimendi, J.A. *Polymer* **23**, 211 (1982).
57. Amass, A.J. and Tuck C.N. *Eur. Polym. J.* **14**, 817 (1978).
58. Ivin, K.J., O'Donnell, J.H., Rooney, J.J. and Stewart, C.D. *Makromol. Chem.* **180**, 1975 (1979).
59. Ivin, K.J., Lapiensis, G., Rooney, J.J. and Stewart, C.D. *J. Molec. Catal.* **8**, 203 (1980).
60. Green, R.M.E., Hamilton, J.G., Ivin, K.J. and Rooney, J.J. *Makromol. Chem.* **187**, 619 (1986).
61. Ho, H.T., Reddy, B.S.R. and Rooney, J.J. *J. Chem. Soc., Faraday Trans. 1* **78**, 3307 (1982).
62. Dainton, F.S. and Ivin, K.J. *Quart. Rev.* **12**, 61 (1958).

4 Transformation reactions
M.J. STEWART

4.1 Introduction

In the previous chapters of this book a number of the newer methods for the synthesis of speciality polymers and copolymers have been discussed. These techniques, which are complementary to the now well established more traditional polymerization techniques of addition (anionic, cationic, radical and coordination) and stepwise polymerizations, are all limited in that none is a universal method and each can only be used with a relatively small number of appropriate monomers. Consequently only certain classes of monomer can be sequentially polymerized by any particular technique, so the range of AB and ABA block copolymers which may be produced is limited. This chapter does not discuss new polymerization procedures which may increase this range, but instead describes the use of novel termination/reinitiation chemistry, generically known as transformation reactions, which enables different polymerization techniques to be used sequentially for the production of a vastly enlarged range of AB and ABA block copolymers.

The synthesis of linear block copolymers, whether by sequential monomer addition, or chain coupling processes, has yielded a variety of materials which have considerable academic interest, as well as commercial importance.[1-2] The first success in the preparation of such materials probably occurred as early as the 1940s with Beyer's work on the cross-linking of polyesters and poly(ester-amides) with di-isocyanates to yield, amongst other products, a series of linear block copolymer elastomers.[3] Other early work, like that of Lundsted, developed low molecular weight poly(oxyethylene-oxypropylene) block copolymers as non-ionic surfactants.[4] Overall, however, the preparation of both of these materials lacked the synthetic control essential for the development of reproducible high molecular weight block copolymer structures necessary for theoretical study and more widespread practical use. Consequently, although both procedures are still used, they have been superseded by more versatile techniques in all but a few extremely specific applications.

It was the discovery by Szwarc of living anionic polymerizations[5-6] which precipitated this change and provided the impetus and procedures for the synthesis of highly ordered linear block copolymers. In such polymerizations, under closely controlled conditions, certain monomers polymerize without transfer or termination reactions occurring; consequently the propagating macromolecule grows proportionately to the amount of monomer introduced. Furthermore, in most cases initiation is fast relative to propagation, so that the molecular weight distribution is extremely narrow. For example, polydispersities ($\bar{M}_w/\bar{M}_n$) of 1.05 have been routinely reported for certain organolithium initiators. Since there is no inherent termination step in such polymerizations, it is also possible to undertake specific termination and transfer reactions on the reactive living polymer intermediates. Thus the preparation of various functional oligomers and polymers containing functionality uniquely at the chain ends is possible. There is not the space, nor need, in this chapter to discuss this technique in any detail, since its physical chemistry has already received excellent coverage from Szwarc[7,8] whilst the synthetic implications of both initiation and termination have been detailed in recent reviews by Fetters[9] and Fontanille.[10]

To prepare block copolymers using living anionic polymerizations, a sequential polymerization synthesis must be used. Thus to prepare an AB block copolymer, the first monomer is converted to its corresponding polymer by use of a monofunctional initiator. This living polymer then serves as an initiator for the second monomer yielding the required AB block copolymer. If required, this may then be used as an initiator for a further aliquot of the first monomer to produce ABA block copolymers. This sequential polymerization technique is amenable to considerable elaboration. For example, BAB blocks may be obtained by use of a bifunctional initiator for the first monomer, and (AB) star polymers by use of multifunctional initiators. Likewise, ABC block copolymers can easily be prepared by using AB block copolymer as the initiator for a third and different segment. The range of possibilities can be extended still further by use of chain coupling reagents, since various forms of star, ladder and three-dimensional matrices can easily be prepared using different functionality initiator and coupling agents. Consequently it is generally accepted that preparing block copolymers by an anionic living polymer mechanism is the best way of generating monodisperse copolymers, which are free of homopolymer impurities, possessing well-defined and predetermined structures.

However, despite the considerable attractions of the anionic living polymer technique, there are also two significant drawbacks. Firstly, it is limited to certain nucleophilic monomers amenable to this techique, such as styrene or butadiene, and excludes all other monomers which polymerize using different active centres. The second and lesser drawback is that, although in theory a wide range of block copolymers could be synthesized, in practice the range is limited owing to the difference in anion reactivities. For example, living polystyrene will initiate the polymerization of ethylene oxide, but the alkoxide ion generated by such a process is more stable than its carbanion predecessor, thus it is not reactive enough to reinitiate styrene polymerization.[11] Consequently the combination of monomers which may be linked experimentally is limited and alternative strategies are required if the preparation of a wide range of block copolymers is to be attempted.

Some progress has been made in the development of alternative living polymerization systems. For example, it has been known for many years that certain cyclic ethers such as tetrahydrofuran (THF) polymerize and copolymerize without chain transfer by a living cationic oxonium ion polymerization process.[12] Unfortunately the range of monomers which work in these conditions is too small for a significant selection of block copolymers to be developed. In Chapter 2 of this book, the development of living group transfer polymerization is discussed in some detail. There is no further need to discuss this mechanism, in either of its forms, apart from noting that it is a powerful tool for the preparation of functional acrylates and methacrylates and, as has been shown, can also be used to prepare linear block copolymers.

More recently, within the last five years, the living carbocationic polymerization of vinyl ethers by use of hydrogen iodide/iodine initiators has been reported by Higashimura and his coworkers.[13-15] (see Section 1.3.4). The mechanism was investigated and found not to include an inherent termination step so that, like anionic polymerizations, functional termination can be undertaken in order to introduce a wide variety of terminal functionality. It has also been shown that, by sequential monomer addition, block copolymers of two different vinyl ethers can be prepared.[16] Another recent advance has been the development of the living carbocationic polymerization of isobutene by Kennedy.[17] Described as a living system, it employs an activated ester polymerization process, which although not suitable for the synthesis of terminally functional polymers, has been reported to give block copolymers with substituted styrenes and isobutene.[18]

Whilst these newer techniques expand the range of living polymerizations somewhat, they still allow access only to the few copolymer structures amenable to preparation by their selected polymerization mode. They therefore still suffer from the problems described previously for anionic systems, in that the range of monomers for any one mechanism is limited and the varying reactivities of the chain ends can prevent certain copolymerizations, especially in the preparation of terpolymers. Consequently new strategies are required if a wide range of linear block copolymers is to be synthesized.

Some such work has been reported in Chapter 6, whereby the synthesis of block copolymers by use of telechelic polymers and macromers has been discussed. Using such techniques a wide range of novel and complex polymeric structures such as dendrimers and arborals can be produced (see Section 1.4.4). However, although these techniques are extremely powerful and greatly extend the range of potential block copolymer morphologies, they lack the fine control of structure and molecular weight associated with true living polymerization systems. These techniques are also ill-suited to the synthesis of reproducible linear or star-block copolymers, materials of interest both in their own right, as well as models for the more complicated molecular architectures.

An alternative approach which, in principle at least, should greatly extend the range of possible monomer combinations in linear block copolymers is to devise a process, or processes, whereby the polymerization mechanism can be changed. In this way vastly different monomers, or groups of monomers, can be polymerized sequentially. Indeed such techniques, which are complementary to the previously described work on telechelic polymers and macromers, have been developed. Although the chemistry of individual reactions differ, they collectively follow a common concept and the description 'transformation reactions' was coined as a generic title for these processes.[19]

4.2 Historical development of transformation reactions

Whilst some reactions which are now classified under the heading of transformation reactions have been known for many years, the unifying framework which pulled all the disparate strands of research together was not postulated until the mid to late 1970s. At this time Richards and his coworkers at PERME (now RARDE) Waltham Abbey, who were researching new routes for the synthesis of novel block copolymers, postulated that conversion between any of the three basic addition polymerization modes, anionic, cationic and free radical, was effectively some form of electron transfer process (Scheme 4.1).[20]

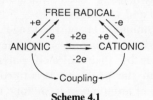

Scheme 4.1

Furthermore interconversion between such modes could be achieved by use of a three step process.

1. Polymerization of monomer (1) by use of polymerization mechanism A and then end-capping the propagating end with a stable but potentially reactive functional group.
2. Isolation of the polymer produced in step 1 above, then dissolving it in a solvent suitable for mechanism B and adding monomer (2).
3. Reaction of the terminal functional group such that it is transformed into a macro-initiator, which causes polymerization of monomer (2) by a different mechanism to that used in the first step, to produce a linear block copolymer.

The methods by which this could be implemented experimentally are now known as transformation reactions. This term is used to describe not just the three simple polymerization mechanisms, but also the range of technologies whereby any addition polymerization mechanism can be converted to another.

In order to simplify the synthetic complexities associated with this three step process, early work concentrated on transformations involving living polymers such as anionic to cationic,[19] since this facilitated the production of idealized polymer structures. Soon after this work commenced, it became apparent that, at least for anion to cation transformations, potentially similar block copolymer structures could be achieved by a direct anion to cation coupling reaction.[21,22] Here the two electron exchange necessary for block copolymer formation is accomplished by direct reaction of a living anion with a living cation, equation 4.1.

$$\sim\sim\sim M_1^- Na^+ + Y^- M_2^+ \sim\sim\sim \longrightarrow \sim\sim\sim M_1 M_2 \sim\sim\sim + NaY \qquad (4.1)$$

Whilst not a true transformation process, since sequential polymerization does not occur, it does in this instance produce similar materials, and it achieves the electron transfer by direct rather than indirect means. Consequently such reactions are normally considered complementary to the three-step transformation process, and will also be discussed in detail later.

As work in this area progressed, the range and scope of transformation reaction technology was expanded. For example, Richards and his various coworkers published routes for the transformation of anionic polymerization to free radical[23] or Ziegler–Natta[24] mechanisms, as well as cationic to anionic[25] and cation to free radical procedures.[26] It is important to note that, at this stage of development, wherever possible at least one, and normally the first, block of the sequence was produced by a living polymerization. Consequently it was easier to study these transformation reactions because, in most cases, the reinitiation chemistry was far from ideal. Thus, if the first segment was prepared in a controlled and reproducible manner, then the study of the functionality and how it behaved when subjected to a transformation process was greatly simplified.

The usual mode of undertaking a transformation reaction, such as those typified by Richards *et al.*, involves the use of a living polymerization which is functionally terminated in near quantitative efficiency by use of a specific terminant. This functional polymer is then used as the basis for a macro-initiator which is incorporated to give a block copolymer. Although a highly specific means of generating terminal functionality, it is not exclusive. In Chapter 6, other methods for the production of telechelic polymers, such as the use of functional initiators or transfer reagents, are discussed. In principle, any of these methods could be adapted for use in transformation reactions and indeed some of the most recent progress in this field has come from within this area. For example, such techniques offer acceptable routes for the transformation of free radical polymerizations into other propagation modes. They consequently represent a considerable expansion to the range of possible transformation reactions. On the debit side, such reactions do suffer from one minor drawback in that, as the chemistry moves away from living or quasi-living systems, the degree of control which can be exerted over the molecular architecture is diminished. This, however, is a small price to pay for the increase in accessible block copolymer morphologies.

In order to demonstrate the full scope of this chemistry, we will now consider the applications of transformation reactions in more detail. In order to simplify matters and illustrate fully the range of chemistry possible, the various transformations have been split into generic classes, these being governed according to the transformation being discussed. In the first section we will consider some of the classical reactions, typified by the work of Richards, involving transformation from anion to cation and vice versa, as well as anion–cation coupling reactions. The subsequent section will discuss the transformation of ionic polymerizations to free radical, as well as some of the potential for radical to ionic reactions. The final section will relate work undertaken on transformations involving at least one polymerization mechanism other than the ionic and radical mechanisms already considered.

4.3 Transformations between anionic and cationic polymerization

4.3.1 *Anion to cation transformations*

This reaction was historically the first specific example of a transformation reaction to be reported as such in the literature and perhaps is one of the most studied and best understood examples of this technology to date. Because of this extensive study it is also probably one of the most efficient examples of transformation technology. The research has undoubtedly been aided by the ability of both anionic and cationic blocks to be prepared under living polymerization conditions, although cationic monomers other than those displaying living polymerization characteristics have also been successfully employed. It has principally been studied using polystyrene as the anionic block, although polybutadiene, polyisoprene and poly(α-methylstyrene) have also been employed with some success. For the cationic segment, polytetrahydrofuran (polyTHF) is the preferred polymer, but various oxetanes along with caprolactam and N-vinyl carbazole have been used also.

As was discussed in the previous section, this particular transformation is effected by changing the initial propagating carbanion to a carbenium ion by a two electron oxidation process. This is achieved experimentally by either a direct coupling between both living anionic and cationic species, or by a three-step transformation process (equation 4.2).

$$\text{\tiny{Termination}} \qquad \text{\tiny{Re-initiation}}$$
$$\sim\!\!\sim\!\!\sim M_1^- Na^+ + RX \longrightarrow \sim\!\!\sim\!\!\sim M_1 R + NaX \longrightarrow \sim\!\!\sim\!\!\sim M_1 R^+ \qquad (4.2)$$

In this particular section we will consider only true anion to cation transformation reactions, which involve end-capping of the living anionic polymer by a reagent possessing a good leaving group such as a halogen, with the associated anion to cation coupling reactions being covered later in Section 4.3.3.

Our starting point is the original work by Richards whereby he introduced the whole concept of transformation reactions and the technique by which they may be undertaken. The specific methodology developed involved termination of the anionic polymerization phase by reaction with bromine, followed by halogen removal, leading to cationic propagation. Subsequently we shall then pass onto the many variations in both reagents and conditions which have been investigated, some of which offer improvements to the original process. Finally, one or two examples of potentially useful new materials which may be prepared by this generic route will be discussed.

Initially this work developed out of previous studies on the functional termination of polystyrenes. In such studies bromine terminated polymers had been prepared by direct reaction of elementary bromine with the living species (equation 4.3).[19] After isolation and purification of this bromine terminated material it was treated with a tetrahydrofuran solution of a silver salt (perchlorate or hexafluorophosphate), to generate the carbocationic initiating species (equation 4.4) which then reacted rapidly with tetrahydrofuran to give a living oxonium ion polymerization, along with inert silver bromide.

$$\sim\sim\sim M_1^- Li^+ + Br_2 \longrightarrow \sim\sim\sim M_1 Br + LiBr \qquad (4.3)$$

$$\sim\sim\sim M_1 Br + AgClO_4 \longrightarrow \sim\sim\sim M_1^+ ClO_4^- + AgBr \qquad (4.4)$$

In these early experiments the overall conversion of polystyrene to block copolymer, when measured by dual detector GPC, did not exceed 60%,[19] in spite of the fact that termination with bromine gave a benzylic bromide species, a species known from model experiments to be a highly efficient cationic co-initiator in conjunction with silver perchlorate or silver hexafluorophosphate. After prolonged examination the problem was diagnosed to lie with the overall efficiency of halogen functionalization since at best only 60–70% bromo-terminated polymer was isolated from the original reaction. The major contaminant was shown to be the Wurtz coupled product (equation 4.5) and its formation was virtually independent of the degree of excess of bromine employed in the reaction. It was therefore concluded that extremely rapid reaction occurred at the polymer-terminant interface before mixing was effected.

$$\sim\sim\sim M_1^- Li^+ + BrM_1 \sim\sim\sim \longrightarrow \sim\sim\sim M_1 - M_1 \sim\sim\sim + LiBr \qquad (4.5)$$

In order to lower this reactivity and consequently avoid the competing Wurtz coupling reaction, it proved necessary to adopt the use of a polymeric Grignard intermediate.[27] This was achieved by reaction of the living polymer

$$\text{\textasciitilde\textasciitilde\textasciitilde} M_1^-Li^+ + MgBr_2 \longrightarrow \text{\textasciitilde\textasciitilde\textasciitilde} M_1MgBr + LiBr \qquad (4.6)$$

with anhydrous magnesium bromide in tetrahydrofuran (equation 4.6). Reaction of the Grignard intermediate with excess bromine was found to be far slower and the desired bromo-functionalized polystyrene was thus produced in yields of up to 95%.[28] Once isolated and purified this bromo-terminated polymer was reacted with silver hexafluorophosphate in bulk tetrahydrofuran at $-10\,°C$, conditions under which living cationic polymerization has been demonstrated to occur and the AB polystyrene-polytetrahydrofuran block copolymer was generated in up to 80% efficiency. Extensive studies of this reaction and its products, primarily by ^1H and ^{13}C NMR, eventually established that the secondary reaction which prevented quantitative block copolymer formation was β-hydride transfer (equation 4.7), which generated a terminal styryl olefinic group. Whilst this reaction could be minimized by use of very low reaction temperatures, quantitative conversion was never achieved. However, in these reactions the copolymer that was actually formed was of extremely narrow polydispersity with values of $\bar{M}_w/\bar{M}_n$ as low as 1.05 being routinely obtained.[28-30]

$$\text{\textasciitilde\textasciitilde\textasciitilde}CH_2-\underset{Ph}{CHBr} + AgPF_6 \longrightarrow \text{\textasciitilde\textasciitilde\textasciitilde}CH_2-\underset{Ph}{CH^+}PF_6^- + AgBr \longrightarrow \text{\textasciitilde\textasciitilde\textasciitilde}CH=\underset{Ph}{CH} + HPF_6$$

(4.7)

At this stage another attempt was made to overcome the β-hydride transfer problem by changing the terminant from elemental bromine to m-xylylene dibromide (1). With this reagent, in which the β-carbon possesses no hydrogen, β-hydride transfer is impossible, thus this side reaction cannot occur. Initially, low functionalization values of about 30% were obtained by direct reaction of the terminant with the living polymer, but these were vastly improved by resort to the Grignard intermediate.[29] Analysis of the termination step by GPC and NMR indicated that use of this reagent, even under optimum conditions, did not give functional termination in more than 80% yield, the product always being contaminated with xylylene coupled polystyrene of twice its molecular weight.[30] This was unfortunate since treatment of this bromoxylyl-functionalized polymer with $AgPF_6$ in tetrahydrofuran at $10\,°C$ was shown to give quantitative conversion to the requisite block copolymer. Thus, by adopting this variation, overall conversion of about 80% was again achieved, so that there was no improvement in conversion compared to the original technique.

(1) m-xylylene dibromide: benzene ring with CH_2Br groups at 1,3-positions.

These techniques have also been successfully used to prepare ABA block copolymers in equally high yield by reaction of the living cationic polytetrahydrofuran species with a difunctional coupling reagent such as the disodium salt of resorcinol (equation 4.8).[20] In these systems the isolation of

$$2\sim M_2^+PF_6^- + NaO\text{-}C_6H_4\text{-}ONa \longrightarrow \sim M_2\text{-}O\text{-}C_6H_4\text{-}O\text{-}M_2\sim + 2NaPF_6 \quad (4.8)$$

the pure polymer is reported to be easier since there is an appreciable difference in product and contaminant solubilities. The styrene contaminant can therefore be removed by simple reprecipitation. It is indeed unfortunate that at the time this investigation was undertaken, a systematic study of this ABA block copolymer was not carried out since in recent years considerable commercial interest has focused on such thermoplastic elastomeric materials. These materials, which still have styrene as their hard segment, may possess superior properties to the known poly(styrene-butadiene-styrene) thermoplastic elastomers by virtue of their polar polyether soft segment, which can wet or interact with a variety of fillers, thus improving the overall performance and physical properties of the composite structure. At present considerable effort is being devoted to developing improvements in this chemistry in order to generate usable bulk materials for an evaluation programme.

These routes and their variations developed by Richards can be considered to be a successful demonstration of the principles of anion to cation transformation. Although the reactions proceed in high efficiency, the reactions are not quantitative and the final product is always contaminated with small amounts of polystyrene homopolymer which can prove difficult to remove, even by fractional precipitation. Scope for further improvement therefore exists and other workers have proposed many techniques in order to produce these block copolymers, but to date, only one other major route has been fully investigated.

This process, which was developed by Franta et al.,[31] involves functionalization of the living polystyrene by phosgene to generate an acyl chloride terminal group (equation 4.9). This is then reacted in a manner similar to that previously described with silver hexafluorophosphate or silver perchlorate to generate an oxocarbenium which then initiates cationic polymerization (equation 4.10). This intermediate oxocarbenium ion is known

$$\sim M_1^-Li^+ + COCl_2 \longrightarrow \sim M_1COCl + LiCl \quad (4.9)$$

$$\sim M_1COCl + AgPF_6 \longrightarrow \sim M_1CO^+ + AgCl \quad (4.10)$$

to be more stable than a normal carbenium ion owing to resonance stabilization (2) so reaction during this phase of the process is clean and fast.

$$\text{R}-\overset{+}{\text{C}}=\text{O} \rightleftharpoons \text{R}-\text{C}\equiv\overset{+}{\text{O}}$$

(2)

Problems, however, arise during the preceding phosgene termination step, which has not yet been made quantitative,[31] although it seems likely that resort to a Grignard intermediate could overcome this problem. This route therefore offers the potential to be the first anion to cation transformation reaction which proceeds in a totally quantitative manner and as such potentially offers a viable route to the routine synthesis of AB, ABA and BAB block copolymers containing anionically and cationically polymerizable segments.

4.3.2 Cation to anion transformations

This transformation is the reverse of the previously discussed anion to cation mechanism. It involves a two-electron addition to the cationic propagating species to give an anionic centre. Several different approaches to this transformation involving direct and indirect techniques have been attempted but, as before, in this section only those methods of indirect transformation involving isolable intermediates will be discussed. Like the aforementioned anion to cation transformations, the most studied materials involve polymers which can be prepared using living polymerization techniques such as polytetrahydrofuran and polystyrene. Several other monomers capable of being polymerized anionically, such as substituted styrenes and dienes, have also been used in the generation of the anionic segment, but until very recently only tetrahydrofuran has offered the control and termination efficiencies necessary for use as the initial cationically-prepared polymer.

The first technique which was developed to specifically tackle this cation to anion transformation involved the termination of living polytetrahydrofuran with lithium cinnamate in order to produce a polymer capped with styryl moieties (equation 4.11). In theory this could then be reacted with an alkyl lithium species in order to generate a carbanion which could initiate anionic polymerization. Indeed, preliminary experiments conducted on small molecule analogues had indicated that reaction with n-butyl lithium did indeed occur, albeit slowly, and that the product could act as an initiator for living anionic polymerization.[25] When attempted using polymeric reagents, experimental examination of the termination step soon led to the development of conditions under which this reaction was quantitative. However, when this styryl functionalized polytetrahydrofuran was subsequently reacted with n-butyl lithium in order to generate the reactive carbanion (equation 4.12), overall conversion to block copolymer did not exceed 20%.[25] After further investigation of this reaction, it was concluded that unspecified side reactions were spoiling this transformation so that this particular route was not a viable cation to anion transformation procedure.

$$\text{\textasciitilde}M_2^+PF_6^- + \text{LiO}-CH_2CH{=}CH(Ph) \longrightarrow \text{\textasciitilde}M_2-CH_2CH{=}CH(Ph) + \text{LiPF}_6 \quad (4.11)$$

$$\text{\textasciitilde}M_2-CH_2CH{=}CH(Ph) + nBuLi \longrightarrow \text{\textasciitilde}M_2-CH_2CH(Ph)-CH(nBu)\text{Li}^+ \quad (4.12)$$

In order to overcome these problems, alternative reactions were investigated, with the most promising candidate being evaluated as an alternative route to this transformation. The selected method involved termination of living polytetrahydrofuran with primary amines such as n-butylamine or aniline, in order to generate a terminal secondary amine functionality (equation 4.13).[32] It was then proposed to metallate this secondary amine functionalized polymer in order to generate a secondary nitranion, which would then act as an initiator for subsequent anionic polymerization (equation 4.14).[33]

$$\text{\textasciitilde}O^+\langle\text{THF}\rangle PF_6^- + RNH_2 \longrightarrow \text{\textasciitilde}O(CH_2)_4NRH + HPF_6 \quad (4.13)$$

$$\text{\textasciitilde}O(CH_2)_4NRH + BuLi \longrightarrow \text{\textasciitilde}O(CH_2)_4NR^-Li^+ + BuH \quad (4.14)$$

In practice, the latter half of this synthetic route posed several major problems. Firstly, it proved almost impossible to conduct a reproducible metallation since one reagent or the other was always in excess. Whilst this may not seem to be a major problem, its actual effect was either to underreact the secondary amine functionalized polytetrahydrofuran giving homopolymer impurities, or worse to overreact and leave traces of metallation reagent, this being a potential anionic initiator in its own right. These problems were eventually overcome by developing a heterogeneous technique whereby excess metal could be removed, leaving a solution of the quantitatively converted secondary nitranion functionalized polymer. Unfortunately another problem manifested itself when this solution was used as an initiator for the anionic polymerization of styrene since, when it was reacted with styrene either by direct addition or slow monomer feed, low initiation efficiencies were observed, with the efficiency being maximized at 30%[33] using a reaction temperature of $-40\,°C$.

The authors of this work argued that this low initiation efficiency was primarily due to the high stability of the secondary nitranion species which makes it a slow initiator for anionic polymerization. They based this assumption on the much broadened GPC traces they observed, which they felt were primarily caused by this slow initiation step. However, they failed to analyse the unreacted polytetrahydrofuran in order to verify that side reactions were not occurring and it now seems likely that competing reactions were also taking place, to the detriment of block copolymer formation. Overall, therefore, this route seems little better than that originally developed, so that the likelihood of synthesizing block copolymers as the major product by either of these routes seems slim.

Thus, the preferred route for the production of block copolymers containing anionically and cationically prepared segments is one involving an anion to cation transformation step. At the present stage of development, the reverse transformation is unreliable and yields block copolymers heavily contaminated with homopolymer impurities. Therefore, until a new cation to anion transformation process is developed, it seems sensible that this route should be avoided.

4.3.3 Anion to cation coupling reactions

So far we have considered only the indirect routes whereby anion can be transformed to cation and vice versa. The most direct way in which block copolymers containing segments to anionically and cationically polymerized monomer can be prepared is by use of a coupling reaction. This mechanism, whilst not conforming to our earlier definition of a transformation process, is usually the simplest technique for the synthesis of these specific types of block copolymer. However, if this chemistry is taken to its logical conclusion, it too becomes a form of transformation process since if a monofunctional polymer (A), prepared using either anionic or cationic techniques, is reacted in equimolar amounts with a difunctional oligomeric species (B), prepared using the polymerization process not used for (A), then coupling can be optimized so that it takes place at only one end of the difunctional oligomer, leaving the other unreacted in order that it can act as an initiator for further propagation.

Richards and his coworkers again were the first to evaluate this technique as an alternative to some of their more traditional transformation processes.[34,35] They undertook extensive evaluation work on these processes using the common and previously described living ionic polymerization systems. Primarily, their work concentrated on using difunctional anionic species, since routes to these dicarbanionic species were well established by this time, whereas routes to difunctional cationic species were less well documented. However, it is important to realize that recent advances in the synthesis of living cationic difunctional oligomers mean that this problem can now largely be overcome.

It is not surprising, in view of the well known nature of the co-reactants, that the reaction between monofunctional anionically prepared polystyrene and monofunctional cationically prepared polytetrahydrofuran was studied first.[34] Experimentation suggested that such a coupling was efficient, provided that reaction stoichiometries were taken into account. Detailed studies conducted on this reaction using ^{1}H NMR and dual detector GPC were then undertaken, and demonstrated that the formation of the expected AB poly(styrene-b-tetrahydrofuran) product was virtually quantitative (equation 4.15).

$$\text{\textasciitilde}CH_2-\underset{Ph}{CH}\text{-}Li^+ + \underset{}{\overset{PF_6^-}{\bigcirc}}\text{\textasciitilde} \longrightarrow \text{\textasciitilde}CH_2-\underset{Ph}{CH}-(CH_2)_4O\text{\textasciitilde} + LiPF_6 \quad (4.15)$$

Subsequent work on the preparation of ABA, BAB and $(AB)_n$ materials using one or more difunctional reagents demonstrated that these reactions also proceeded with high efficiency, although small excesses of either anionic or cationic reagent prevented quantitative conversion to the requisite block copolymer. Consequently, despite requiring precise estimation of reagent concentration, this route is perhaps the preferred route for the laboratory synthesis of poly(styrene-b-tetrahydrofuran) copolymers because of its higher yield relative to existing transformation reactions. However, on a larger scale, the matching of stoichiometries has proved almost impossible, so that one reagent or other is in excess and block copolymers with appreciable homopolymer contamination are prepared. This problem is one inherent in such ionic coupling reactions and, when undertaken on a larger scale, is difficult to overcome. It is therefore likely that, for larger scale preparation of these materials, this route will not be employed, with alternative methods based on new developments in transformation technology being preferred.

Attempts to extend this technology to the incorporation of other anionically polymerizable monomers were less successful. When α-methylstyrene or butadiene[35] was used as the anionically polymerizable monomer, it was observed that coupling was not the major reaction and less than 30% block copolymer formation was obtained. Whilst the competing side reaction with polybutadiene has not been adequately described, considerable effort was expended by the authors in order to determine the cause of this problem. It was found that coupling does not readily occur with living poly(α-methylstyrene), which causes the formation of terminal olefinic groups on the polytetrahydrofuran chain and proton terminated poly(α-methylstyrene).[35] A more recent publication by Kucera indicates that this side reaction can be overcome and that coupling between the disodium(α-methylstyrene)tetramer and polyTHF proceeds in high efficiency so that block copolymers are the main product. However, there is no absolute indication of the transformation efficiency in this work, and homopolymer contamination is still a problem.[36]

It is clear therefore that, whilst this may be the technique of choice when employing polystyrene and polytetrahydrofuran, proton transfer side reactions can cause unacceptable homopolymer contamination when polybutadiene, or possibly poly(α-methylstyrene), are employed. In such circumstances other methods of producing these materials may be more successful; these include the previously described indirect transformation reactions, as well as indirect coupling reactions whereby one living polymer is reacted with a functionally terminated polymer prepared using the opposite polymerization technique. Using this latter technique, a range of novel poly(butadiene-b-tetrahydrofuran) and poly(α-methylstyrene-b-tetrahydrofuran) copolymers have recently been produced.[80]

4.4 Transformations between ionic and free radical polymerization

4.4.1 *Anion to radical transformations*

The anion to free radical transformation process is the most widely investigated application of this technology. Several viable routes have been extensively investigated by various authors from around the world and consequently a wide range of materials developed. Richards and his collaborators were not the earliest workers active within this particular segment of transformation reaction technology and indeed significant work was undertaken by various parties before the term transformation reaction was even coined. This transformation involves a single electron transfer process, which is normally achieved by functional termination of the anionic segment by a reagent which introduces a radical precursor. This precursor is then heated or irradiated in order to generate polymeric radicals, which may then initiate polymerization.

The variety of molecules which may be produced by this process is potentially vast and covers nearly all possible linear block copolymer morphologies, as well as a vast range of monomer pairs. By use of a monofunctional anionic segment (A) it is possible to synthesize both AB or ABA copolymers if the radical polymerization phase (B) terminates by disproportionation or combination, respectively. Likewise, use of a difunctional anionic segment allows BAB and $(AB)_n$ copolymers to be produced by radical disproportionation and combination. When this versatility is coupled with the wide range of monomers which can be utilized for both anionic (styrenes and dienes) and free radical (acrylates, methacrylates, acrylonitriles and many other vinyl monomers) reactions, the range of potential block copolymers accessible by this route is large. In this particular section only the major routes and typical products obtained with common monomers will be discussed in any detail. The potential for the synthesis of other block copolymer systems, especially involving variations in the radically polymerized segment is, however, evident and is left to the reader to explore.

The early approaches to the change of the propagating centre from anionic to free radical fall into two main categories. These involve functionalization with oxygen,[37,38] or termination with azido[39-41] or peroxy-species.[42] Considerable effort was expended in these studies and it was indeed unfortunate that this early work was not conducted with the support of modern analytical facilities, since, in the absence of experimental details on the conversion, the products are often inadequate. Consequently this makes the evaluation of these earlier experiments difficult, especially with respect to claimed efficiencies of transformation. Subsequent and more recent work by Richards and others[23,43] has developed this theme and generated more potentially specific techniques and clearly demonstrated, with the aid of NMR and GPC, the efficiency and degree of conversion. Thus evaluation of these latter transformation processes is comparatively easy. Direct comparison between the early and these later techniques is impossible, so all major methods have been included for completeness. It is left to the reader to decide which route offers the best potential for a particular system, although it should be noted that, as a general rule, the later the publication, the more repeatable the results.

The earliest work in this field was recorded by Szwarc who in 1956 reacted living polystyrene with oxygen in order to generate the hydroperoxide (equation 4.16). He also observed a coupled polystyrene product of twice the molecular weight.[5] It was reasoned that this coupled product was probably the result of an electron transfer process to oxygen which yielded a polystyryl radical and an oxygen radical anion. The polystyryl radicals then rapidly combined to give the high molecular weight material (equation 4.17). At this time, the science of living polymers was still in its infancy so no attempt was made to evaluate the efficiency of this transfer process, or to use it in the production of block copolymers, although in this work Szwarc did indicate that this coupled material was a relatively minor product of the reaction.

$$\sim\sim CH_2-\underset{Ph}{CH}Li^+ + O_2 \longrightarrow \sim\sim CH_2-\underset{Ph}{CH}-O-O^-Li^+ \xrightarrow{H^+} \sim\sim CH_2-\underset{Ph}{CH}-O-OH + Li^+ \quad (4.16)$$

$$\sim\sim CH_2-\underset{Ph}{CH}Li^+ + O_2 \longrightarrow \sim\sim CH_2-\underset{Ph}{CH}\cdot + O_2^- \longrightarrow \sim\sim CH_2-\underset{Ph}{CH}-\underset{Ph}{CH}-CH_2\sim\sim \quad (4.17)$$

Later workers have also attempted to use this oxygenation technique but, unlike Szwarc, attempted to use the major component, the polymeric hydroperoxide, as their transformation reagent. Formation of this species can be increased relative to the coupled product by use of excess oxygen, as the excellent investigations on this subject by Brossas revealed.[37,38] In these reports, the generation of free radicals by the reaction of the polymeric hydroperoxide species with ferrous sulphate (equation 4.18), or by heating the polymer to 150 °C, was demonstrated (equation 4.19). It was perhaps unfortunate that although block copolymer was formed during these reactions in significant amounts, it was not the major product. This is especially true in the latter case where a hydroxyl radical, which can also initiate homopolymerization, was produced by the same reaction. Consequently transformation efficiency is far from quantitative.

$$\text{\textasciitilde\textasciitilde}CH_2-\underset{\underset{Ph}{|}}{CH}-O-OH \xrightarrow{Fe^{2+}} \text{\textasciitilde\textasciitilde}CH_2-\underset{\underset{Ph}{|}}{CH}-O \cdot + Fe^{3+} + HO^- \qquad (4.18)$$

$$\text{\textasciitilde\textasciitilde}CH_2-\underset{\underset{Ph}{|}}{CH}-O-OH \xrightarrow{\Delta} \text{\textasciitilde\textasciitilde}CH_2-\underset{\underset{Ph}{|}}{CH}-O \cdot + HO \cdot \qquad (4.19)$$

The other major approach to this transformation was developed in the early 1970s and involved the introduction of free radical initiators as terminal groups on the anionic segment. The first of these developments involved the introduction of azo groups into the polymer by reacting the living anionic polymer with 2,2′-azobisisobutyronitrile (AIBN) to give potentially two different coupled polymeric products with a central azo moiety (Scheme 4.2).[39,40] Investigation of this termination reaction by use of ^{14}C labelled nitrile groups on the AIBN molecule conclusively demonstrated that the major reaction was that of nucleophilic displacement of the nitrile groups.[41] Furthermore it was demonstrated also that this reaction could be made virtually quantitative by varying the conditions and by using potassium ion as the gegenion in a polar solvent, such as THF, with reaction temperature kept below 20 °C.

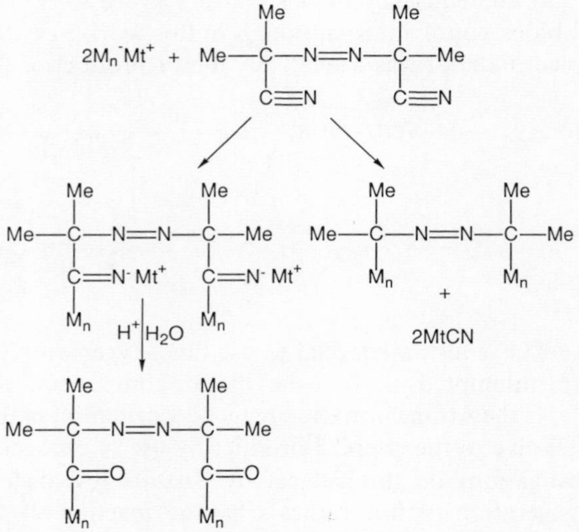

Scheme 4.2

This pure azo linked polymer has been shown successfully to give block copolymers of the AB and ABA types with vinyl chloride, but unfortunately no details of initiation efficiencies, or conversion to block copolymer, were given. Thus it is impossible to estimate whether the degree of conversion was quantitative or of far lower yield. Consequently no definite assessment can be made of the efficiency of this particular transformation.

The other technique which was developed to introduce free radical initiators during the anionic termination phase is conceptually similar to that discussed above and involves the reaction of anionic living polymers with substituted peroxides.[42] For example, living polystyrene has been terminated with 4,4'-bis(halomethyl)benzoyl peroxide to give a coupled polymer with central peroxide functionality. Unfortunately there was significant nucleophilic attack on the perester moiety during the anionic termination step so that, at best, only 50% of the theoretical peroxide was incorporated. This peroxide was then thermally decomposed to yield polymeric alkoxy radicals which successfully copolymerized both methyl methacrylate and vinyl chloride in high efficiency (Scheme 4.3). It is unfortunate that this otherwise successful transformation methodology should be spoilt by unwanted nucleophilic side reactions, although it has been reported that yield can be increased by creating a less reactive carbanion prior to reaction with the peroxide.[43] This may in the future allow this reaction to be developed into a highly efficient transformation process.

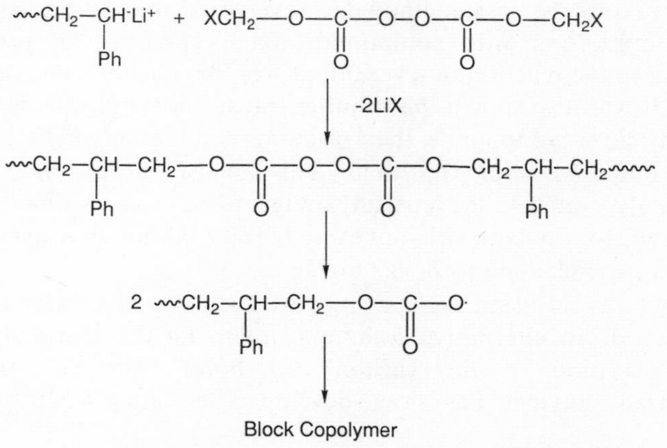

Scheme 4.3

The more recent work by Richards and his coworkers has developed similar themes. The work has concentrated on the synthesis of anionically prepared polymers possessing unstable carbon–metal bonds. Most work has been undertaken by a two-stage process, whereby the living anionic polymer is terminated with a trialkyl lead chloride (equation 4.20), to generate a relatively stable trialkyl lead functionalized polymer.[23,44] This product is then reacted with a soluble silver salt, such as $AgNO_3$, $AgCN$ and $AgClO_4$, in order to form an unstable carbon–silver bond; this then can be used to initiate radical polymerization (equation 4.21).[45]

$$\sim\sim\sim M_1^-Li^+ + ClPbR_3 \longrightarrow \sim\sim\sim M_1PbR_3 + LiCl \quad (4.20)$$

$$\sim\sim\sim M_1PbR_3 + AgY \longrightarrow \sim\sim\sim M_1-Ag + YPbR_3 \longrightarrow \sim\sim\sim M_1\cdot + Ag \quad (4.21)$$

Although block copolymers with such comonomers as methyl methacrylate, styrene, butadiene and acrylic acid have been produced, it is reported that the transformation is not quantitative. This is unfortunate since it has been demonstrated that the addition of trialkyl lead chlorides, especially triethyllead chloride, to the living polymer can be made quantitative,[46] thus the inefficiency must occur during the formation and reaction of the carbon–silver bond. From the experimental results presented by Richards and Schué, it seems likely that the reason why non-quantitative initiation occurs is that the polymeric radicals formed by this procedure possess lifetimes which are far too short to completely initiate the comonomer. Thus side reactions, such as β-proton elimination in polystyrenes, deactivate the radical centre. Certainly, under detailed examination, polystyrenes have been shown to possess far lower efficiencies than polyisoprenes, where reactions such as β-proton elimination cannot occur. It has also proved possible, by using complexing agents such as crown ethers, to improve upon the transformation efficiency of polystyrenes over the basic method. However, care must be exercised, or else strong complexation with compounds like cryptands can prove to be detrimental to the reaction. A few examples of this phenomenon are given in Table 4.1. It was also shown that if other transition metal salts were used in place of the silver compounds, then reaction with the polymeric trialkyllead adduct resulted in improved yield. This is also shown in Table 4.1.[45] Unfortunately, even with the most advantageous method possible, conversion of polystyrene to copolymer has not exceeded 40%, although conversions with polyisoprene are claimed to be far higher.

In order to avoid using the over-reactive carbon–silver complexes of the above method, an alternative technique involving the thermally induced homolytic scission of the carbon–lead bond from the trialkyllead functionalized polymer has been developed (equation 4.22).[46] Potential

Table 4.1 Transformation efficiencies of chemical reactions between transition metal salts and the polystyrene-triethyllead adduct

Salt	Conversion of polystyrene (%)	Conversion of methyl methacrylate (%)
$AgClO_4$	5	5
$AgClO_4$ + 12 crown 4	30	7
$AgClO_4$ + 18 crown 6	15	35
CuOTf	10	32
$Cu(OTf)_2$	23	20
$Cr(NO_3)_2$	34	33

Tf = Triflate

homopolymer contamination from reaction of trialkyllead radicals, also produced by this process, has been shown to be minimal, owing to the rapid combination of such species to give inert products (equation 4.23). Consequently conversion to the block copolymer is claimed to be high, although not quantitative.

$$\sim\sim\sim M_1PbR_3 \xrightarrow{\Delta} \sim\sim\sim M_1 + R_3Pb \xrightarrow{nM_2} \text{Block Copolymer} \qquad (4.22)$$

$$4R_3Pb\cdot \longrightarrow 3R_4Pb + Pb\downarrow \qquad (4.23)$$

A similar thermal method has been developed using mercury salts in place of the lead compounds. In this variation a functionalized polymer is obtained by reacting living anions with an excess of mercuric chloride (equation 4.24). Depending on reagent stoichiometry, either the desired product or a polymeric fraction of twice the molecular weight, was produced as the major product.[47] However, recent work has disputed the earlier claim that the monofunctional polymeric mercury compound could be uniquely isolated and has indicated that mercury-free coupled product will always be present as a homopolymer contaminant.[48] Subsequent thermolysis of the polymeric mercurohalide compound has been shown to yield the required block copolymer, although this is always accompanied by the formation of a significant amount of coupled polymeric product (equation 4.25). Some further improvement was noted when oxidative cleavage with copper (II) bromide, or exchange reactions with mercuric halides were used to initiate homolysis of the carbon–mercury bond, but the overall transformation efficiency remained low.[48]

$$\sim\sim\sim M_1^-Li^+ + HgCl_2 \longrightarrow \sim\sim\sim M_1HgCl + LiCl \qquad (4.24)$$

$$\sim\sim\sim M_1HgCl \xrightarrow{\Delta} \sim\sim\sim M_1 + HgX \xrightarrow{nM_2} \text{Block Copolymer} \qquad (4.25)$$

One final technique, which is possibly also the most successful so far developed to tackle these anion to radical transformations, has been pioneered by Bamford and Eastmond at Liverpool. This method, which is widely applicable (we will also encounter it in cation to radical and group transfer to radical transformaions), involves reaction of a bromine terminated polymer, similar to those already discussed in Section 4.31 for anion to cation transformations (equations 4.3 and 4.6), with the irradiation product of dimanganese decacarbonyl in order to generate radical macroinitiators (equation 4.26).[49,50] In this way a variety of AB and ABA block copolymers with methyl methacrylate, methyl acrylate, ethyl acrylate, isoprene and chloroprene as the second segment have been prepared. The actual structure is dependent, as described earlier in this section, on the mode of radical termination.

This work reaffirmed that bromine termination efficiency was not 100%,

$$\text{\textasciitilde\textasciitilde CH}_2\text{--CHBr} + \text{Mn(0)} \longrightarrow \text{\textasciitilde\textasciitilde CH}_2\text{--CH·} + \text{Mn(I)} \quad (4.26)$$
$$\qquad\quad |\qquad\qquad\qquad\qquad\qquad |$$
$$\qquad\quad \text{Ph}\qquad\qquad\qquad\qquad\quad \text{Ph}$$

Where Mn(0) = Active product from irradiation

and that some of the impurities consequently present retarded the radical copolymerization process. However, if care was taken in purifying the initial bromine containing polymer, then these macroinitiators behaved similarly to low molecular weight halide initiators, and so produced well defined systems. It was also claimed that all bromine containing polymers participated in the reinitiation process so that a virtually quantitative transformation process occurred. Similar conclusions have been drawn by Niwa et al. who also used this technique to make a series of highly reproducible block[51] and graft copolymers.[52] Thus, this reaction, one of the most recent developments in this area of transformation technology, also seems one of the most successful, with typically 70–80% conversion of homopolymer to block copolymer material being obtained, the residue being due to unbrominated side products from the initial termination step.

4.4.2 *Cation to radical transformations*

Although anion to radical transformations have been extensively investigated, there has not been the same degree of interest in the corresponding cation to radical transformation reactions. This is almost entirely due to the lack of living cationic systems which, apart from polyTHF, were almost unknown until very recently. Thus, the wide range of synthetic options open in the termination phase of anionic systems has not been repeated to the same extent for cationic systems. In spite of its apparent neglect, many viable transformation strategies, which in many respects parallel anion to radical developments, have emerged over the years.

The first procedure attempted to introduce a common radical initiator functional group, such as peroxide, into the centre of the polymeric chain so that subsequent thermolysis would generate two polymeric radicals. Burgess successfully achieved this[44] by reacting the sodium salt of succinic acid peroxide with living polyTHF to give a peroxy-coupled polyTHF species (equation 4.27). It was also demonstrated that, in the presence of suitable

$$\text{\textasciitilde\textasciitilde}M_2^+PF_6^- + Na^+\text{-}O\text{---}\underset{\underset{O}{\|}}{C}\text{-}(CH_2)_2\text{-}\underset{\underset{O}{\|}}{C}\text{---}O\text{---}O\text{---}\underset{\underset{O}{\|}}{C}\text{-}(CH_2)_2\text{-}\underset{\underset{O}{\|}}{C}\text{---}O^-Na^+$$

$$\Big\downarrow -2NaPF_6$$

$$\text{\textasciitilde\textasciitilde}M_2\text{-}O\text{---}\underset{\underset{O}{\|}}{C}\text{-}(CH_2)_2\text{-}\underset{\underset{O}{\|}}{C}\text{---}O\text{---}O\text{---}\underset{\underset{O}{\|}}{C}\text{-}(CH_2)_2\text{-}\underset{\underset{O}{\|}}{C}\text{---}O\text{---}M_2\text{\textasciitilde\textasciitilde} \quad (4.27)$$

comonomers, the radical macroinitiator could be generated and AB and ABA block copolymers formed.[44] Unfortunately, a critical appraisal of this system was not undertaken, so that the overall transformation efficiency of this route is unknown. The report does, however, indicate that considerable homopolymer contamination is present in the copolymer products, indicating that such a route is far from quantitative.

$$Cl-\overset{O}{\overset{\|}{C}}-(CH_2)_2-\overset{CH_3}{\underset{CN}{\overset{|}{C}}}-N=N-\overset{CH_3}{\underset{CN}{\overset{|}{C}}}-(CH_2)_2-\overset{O}{\overset{\|}{C}}-Cl \quad (4.28)$$

$$\text{AgBF}_4 \Big| \text{THF}$$

$$BF_4^- \quad \langle O^+ \sim\sim \overset{O}{\overset{\|}{C}}-(CH_2)_2-\overset{CH_3}{\underset{CN}{\overset{|}{C}}}-N=N-\overset{CH_3}{\underset{CN}{\overset{|}{C}}}-(CH_2)_2-\overset{O}{\overset{\|}{C}}\sim\sim^+O \rangle \quad BF_4^-$$

A conceptually similar approach has been developed by Yagci *et al.* whereby polyTHFs containing a central azo linkage have been synthesized by use of a functional initiator (equation 4.28). This central linkage is then decomposed in the presence of comonomer to give block copolymers (equation 4.29).[53,54]

$$\sim\sim\overset{O}{\overset{\|}{C}}-(CH_2)_2-\overset{CH_3}{\underset{CN}{\overset{|}{C}}}-N=N-\overset{CH_3}{\underset{CN}{\overset{|}{C}}}-(CH_2)_2-\overset{O}{\overset{\|}{C}}\sim\sim \xrightarrow{\Delta} \sim\sim\overset{O}{\overset{\|}{C}}-(CH_2)_2-\overset{CH_3}{\underset{CN}{\overset{|}{C}}}\cdot \xrightarrow{nM_2} \text{Block Copolymer}$$

(4.29)

Conversion is reported to be extremely high and relatively independent of cationic block length. Moreover, by use of this functional initiator, living conditions for polyTHF have been established, so that reproducible block lengths can be synthesized.[55] This work thus demonstrates an extremely viable cation to radical transformation process and is a good example of the flexibility of the functional initiator approach, an approach which offers considerable scope for future developments.

Another approach with a similar theme involves terminating living cationic polyTHF with a mercaptan in order to generate and isolate a polymeric mercaptan radical transfer reagent (equation 4.30). Unfortunately, by its very nature, this route also generates inert homopolymer as well as the required block copolymers (equation 4.31). Thus, regardless of transformation efficiency (which is unknown) the product is seriously contaminated with homopolymer byproduct.[43]

$$\sim\sim M_1^+ Y^- + SH_2 \longrightarrow \sim\sim M_1 SH + HY \quad (4.30)$$

$$\sim\sim M_1 SH + R\cdot \longrightarrow \sim\sim M_1 S\cdot + RH \quad (4.31)$$

One of the most successful demonstrations of cation to radical transformation technology to date involves work by Eastmond on a cationic variation of the successful anion to radical process described in Section 4.3.3 above. In this method the polyTHF is terminated by the sodium salt of bromoacetic acid in a virtually quantitative reaction (equation 4.32).[56] This is then reacted with the irradiated product of dimanganese decacarbonyl, in an identical manner to the bromo-terminated polystyrenes of the anionic example (equation 4.26), in order to generate radical macroinitiators. These then readily copolymerize with methyl methacrylate to give block copolymers in high efficiency.[57]

$$\sim\sim\sim M_1^+Y^- + Li^+{}^-O-\overset{O}{\underset{\|}{C}}-CH_2Br \longrightarrow \sim\sim\sim M_1-O-\overset{O}{\underset{\|}{C}}-CH_2Br + LiY \quad (4.32)$$

$$\sim\sim\sim OH + OCN-CCl_3 \longrightarrow \sim\sim\sim O\text{-}OCNH\text{-}CCl_3 \quad (4.33)$$

$$\sim\sim\sim NH_2 + OCN-CCl_3 \longrightarrow \sim\sim\sim NH\cdot OCNH\text{-}CCl_3 \quad (4.34)$$

$$\sim\sim\sim COOH + OCN-CCl_3 \longrightarrow \sim\sim\sim O\text{-}CNH-CCl_3 + CO_2 \quad (4.35)$$

Bamford has recently indicated that reactions of certain functional polymers with halo-isocyanates, although not specifically developed for cation to radical transformations, quantitatively yield bromofunctionalized polymers (equations 4.33–4.35).[58] These materials may then be irradiated in the presence of dimanganese decacarbonyl, in order to generate block copolymers. In an illustrative example with polyethylene glycol (molecular weight 3300–4000) almost complete conversion to the copolymer was achieved. It was reported that haloisocyanates also react rapidly with amines and carboxyls to give functionalized polymers which may then be amenable to copolymerization. Consequently any polymerization methodology which can routinely generate such terminal moieties can be used to generate block copolymers. It therefore seems probable that hydroxyl-functionalized chain ethers produced by cationic procedures can be used to produce block copolymers in high efficiency by this method.

4.4.3 Radical to ionic transformations

There has been a relative lack of interest in the investigation of these particular transformations, primarily because of the short lifetimes of the radical precursors. The main method of achieving a radical to ionic polymerization transformation has centred on the development of a radical to cationic process using redox reagents to functionalize the radicals. For example, the reaction of ferric bromide with polystyryl radicals has been investigated and been shown to yield primarily a polymeric bromine adduct (equation 4.36). This has then been reacted with silver salts such as $AgClO_4$ and $AgPF_6$ to generate a carbocation which is capable of initiating copolymerization (equation 4.37).

$$\text{\textasciitilde\textasciitilde CH}_2\text{—CH}^\cdot\underset{\underset{\text{Ph}}{|}}{} + \text{FeBr}_3 \longrightarrow \text{\textasciitilde\textasciitilde CH}_2\text{—CH—Br}\underset{\underset{\text{Ph}}{|}}{} + \text{FeBr}_2 \quad (4.36)$$

$$\text{\textasciitilde\textasciitilde CH}_2\text{—CH—Br}\underset{\underset{\text{Ph}}{|}}{} + \text{AgY} \longrightarrow \text{\textasciitilde\textasciitilde CH}_2\text{—CH}^+\text{Y}^-\underset{\underset{\text{Ph}}{|}}{} + \text{AgBr} \quad (4.37)$$

An alternative to this reaction involves reaction of the polystyryl radical with carbon tetrabromide in order to synthesize a bromo functionalized polystyrene. This has then been reinitiated using silver salts to give block copolymers with indene, THF, vinyl pyrrolidone and vinyl carbazole. Unfortunately, no details of transformation efficiencies are available.[43]

One recent development has been the publication of preliminary findings on a radical to cation transformation by Yagci et al. This process involves the same functional initiator used for his cationic to radical transformation reactions (Section 4.4.2) except that it is used in the reverse mode. Thus, by conducting the radical polymerization first, chloro-functionalized vinylic polymers result. It is claimed that these can be efficiently reinitiated to polymerize THF to yield AB block copolymers.[81]

In passing, it is interesting to note that Richards has proposed, but not experimentally investigated, a route for a radical to anion transformation.[20] He proposed that the radical polymerization phase of this reaction be undertaken in a non-solvent for the homopolymer. Thus macroradicals will be occluded in the precipitated polymer. This gel could then be swollen in the presence of an appropriate solvent and reducing agent, such as the sodium naphthalene radical anion in THF, to obtain an anion which can initiate block copolymerization. This is illustrated in equation 4.38.

$$\text{\textasciitilde\textasciitilde M}_1\cdot + \text{Na}^+\text{Nap}^- \longrightarrow \text{\textasciitilde\textasciitilde M}_1^-\text{Na}^+ + \text{Nap} \xrightarrow{n\text{M}_2} \text{\textasciitilde\textasciitilde M}_1(\text{M}_2)_n^-\text{Na}^+ \quad (4.38)$$

4.5 Transformations involving other modes of polymerization

4.5.1 Anion to Ziegler–Natta transformations

The transformation from anionic to Ziegler–Natta polymerization mechanism cannot be regarded as a simple electron transfer process, unlike the earlier examples. Consequently a specific methodology has had to be developed for this process. The reaction between aluminium alkyls and titanium (IV) chloride to give Ziegler–Natta polymerization is well known and documented.[59] It involves a two-stage process whereby an aluminium alkyl such as trialkylaluminium is reacted with $TiCl_4$ in order to generate the active $\beta TiCl_3$ for the second phase, whilst a reactive alkyl radical is eliminated. This latter species plays no further part in the reaction because it terminates rapidly

$$TiCl_4 + R_3Al \longrightarrow RTiCl_3 + R_2AlCl \longrightarrow \beta\text{-}TiCl_3 + R^{\cdot}$$

$$\beta\text{-}TiCl_3 + R_3Al \longrightarrow RTiCl_2 + R_2AlCl \xrightarrow{C_2H_4} \begin{array}{l} RTiCl_2C_2H_4 \\ \text{Coordination} \\ \text{or} \\ RCH_2CH_2TiCl_2 \\ \text{Insertion} \end{array}$$

Scheme 4.4

to give inert alkyl products. The second phase is reaction of the active $\beta TiCl_3$ with further trialkylaluminum to generate a species capable of initiating polymerization. This also results in the incorporation of the second alkyl group into the polymer. This two stage process is illustrated in Scheme 4.4. Polymeric aluminum alkyl compounds, which can take the place of small molecule analogues have been successfully synthesized by the alkylation of aluminium halides with living anionic polymer (equation 4.39), to give uniquely mono- di- or tri(polymero)aluminum species.[24]

$$\sim\!\!\sim\!\!M^{-}Li^+ + AlCl_3 \longrightarrow (\sim\!\!\sim\!\!M)_3Al + (\sim\!\!\sim\!\!M)_2AlCl + \sim\!\!\sim\!\!M\text{—}AlCl_2 \quad (4.39)$$

Furthermore it has been reported that all of these (polymero)aluminum species react with titanium (IV) chloride, in the presence of ethylene, to give block copolymers. Different efficiencies, not exceeding 40%, have been reported for the various (polymero)aluminum species depending on whether polyisoprene or polystyrene was used as the living anionic polymer. These results were interpreted in terms of steric effects, which reduced the transformation efficiency of highly alkylated compounds when bulky polystyrene moieties were used as the living anionic polymers. Inert homopolymer contamination is a serious problem in this transformation since not only is unreacted homopolymer present, but material from the formation of polymeric radicals during the first phase reduction reaction also adds to the contamination. Indeed it is likely that this material is the major contaminant. It is perhaps fortunate that, in this case at least, unreacted homopolymer can be removed by an extended extraction process, so that the desired AB block copolymer can be isolated in high yield.[24]

In a series of recent reports, Aldassi[60-62] and Amass[63] have indicated that this method has also been adopted to produce AB poly(styrene-b-acetylene) copolymers. The reported transformation efficiency is extremely low, with only about 5–10% of the required product being isolated. These materials, after rigorous purification, are currently being evaluated for use as electroactive polymers.[64]

4.5.2 *Anion to metathesis transformations*

One of the most recently developed transformation processes is that of anion to metathesis, with the first example being reported by Amass *et al.* in 1985.[65]

TRANSFORMATION REACTIONS

In this work, polystyryllithium was used as a co-initiator with tungsten hexachloride for the polymerization of cyclopentene. It was claimed that polymerization proceeded by incorporation of the polystyrene to produce block copolymers. Although conversion was not quantitative, this and a subsequent study[66] collected enough evidence to propose a novel reaction mechanism for this transformation. This involved the formation of a bimetallic bridged intermediate which subsequently rearranged to form the metathesis active centre. This is shown in Scheme 4.5 below.

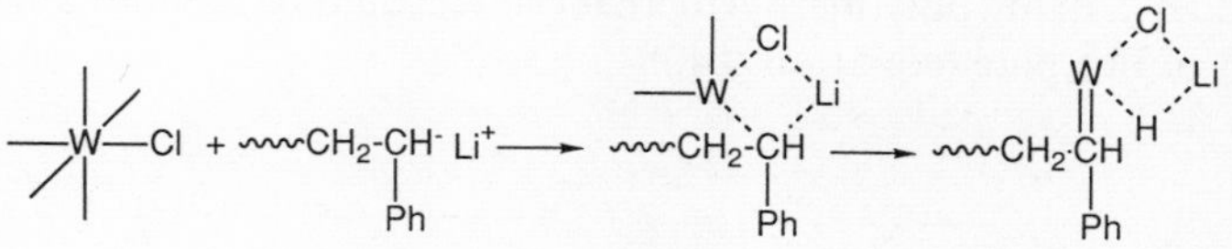

Scheme 4.5

This transformation process is as yet far from optimized, with low conversion and considerable polystyrene homopolymer contaminant being reported. It is, however, likely that further optimization of this anion to metathesis transformation process will significantly increase block copolymer yields. Until such time as an improved method is developed, it was claimed that pure copolymer could be isolated by extensive purification.

4.5.3 Ziegler–Natta to radical transformations

This transformation, which was first reported by Agouri as early as 1972, is used to prepare alkenic–vinylic block copolymers.[67] This was achieved by inducing Ziegler–Natta polymerization using diethylzinc as the transition metal complex, then switching to a radical polymerization, firstly by oxidizing the carbon–zinc bond, then thermolytically cleaving the resultant peroxide. These reactions are outlined in Scheme 4.6.

Although the transformation itself is fairly efficient, it is unfortunate that this method suffers from serious contamination problems. This has been attributed to homopolymer produced from the competing ethoxide radical formation, although side products from the transformation process also cannot be discounted.

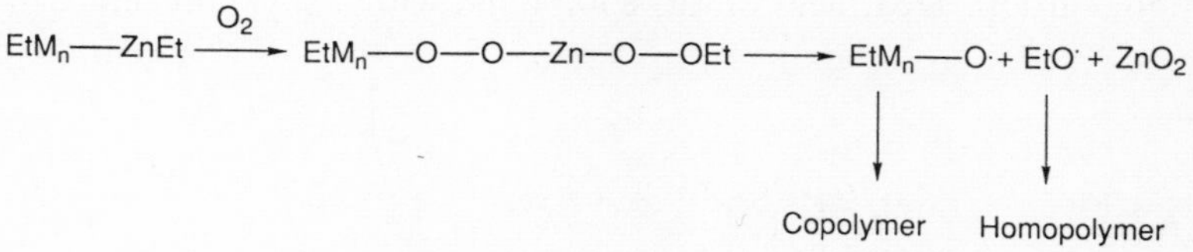

$$EtM_n\text{—}ZnEt \xrightarrow{O_2} EtM_n\text{—}O\text{—}O\text{—}Zn\text{—}O\text{—}OEt \longrightarrow EtM_n\text{—}O\cdot + EtO\cdot + ZnO_2$$

Copolymer   Homopolymer

Scheme 4.6

4.5.4 Group transfer to radical transformations

The next two transformations involve two different types of group transfer polymerization. As this new method of polymerization, developed by DuPont, has already been discussed at length in Chapter 2, it will be assumed that the reader has some familiarity with this polymerization mechanism.

The first of these two examples is of the transformation from a group transfer to a free radical mechanism by use of a bromo-functionalized polymethylmethacrylate. It was reported by workers at DuPont that the active end of a living poly(methyl methacrylate) could be functionalized by the addition of bromine (equation 4.40).

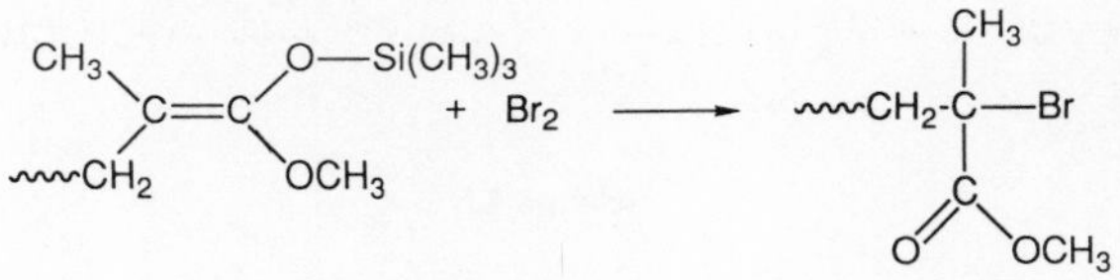

$$\underset{\sim\sim CH_2}{\overset{CH_3}{\diagdown}} C=C \underset{OCH_3}{\overset{O-Si(CH_3)_3}{\diagup}} + Br_2 \longrightarrow \sim\sim CH_2-\underset{\underset{OCH_3}{\overset{\|}{C}}}{\overset{CH_3}{\underset{|}{C}}}-Br \qquad (4.40)$$

Eastmond et al.[68] then employed their previously discussed technique of reacting bromo-functionalized polymers with the product of irradiated dimanganese decacarbonyl (equation 4.26), in the presence of styrene to give a mixture of AB and ABA block copolymers. Significant unreacted poly(methyl methacrylate) was also reported by the authors, so contamination problems exist. However, the authors state that under the conditions used no effort was made to maximize conversion and conclude that overall this procedure is an effective method of synthesizing block copolymers containing narrow polydispersity poly(methyl methacrylate).

4.5.5 Metathesis to aldol-group transfer transformations

The second example concerns the transformation from metathesis to aldol group transfer polymerization so that poly(norbornene-b-vinyl alcohol) copolymers resulted. This work stems from the recent discovery of living ring-opening metathesis polymerization using bis(η^5-cyclopentadienyl)titanocyclobutane compounds,[69,70] (structures 3 and 4) as initiators and norbornene as monomer. The resultant polymer chain, which possesses a terminal cyclobutane moiety, may be functionalized by reaction with terephthaldehyde, an aromatic dialdehyde, to give polymers with an aldehyde functional group. Treatment of this compound with a silyl vinyl ether in the

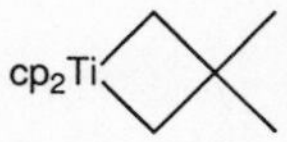

      cp₂Ti⟨▱⟩   cp₂Ti⟨▱⟩

        (3) 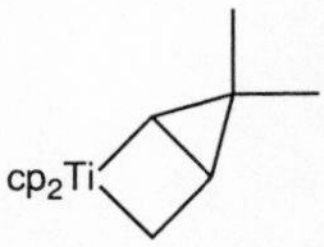   (4)

presence of a Lewis acid, causes silyl group migration, with the simultaneous generation of a terminal aldehyde moiety.[71] On termination, the silyl groups can be removed with tetrabutylammonium fluoride followed by methanol to generate a poly(vinyl alcohol) block copolymer segment.

Transformation efficiency is reported to be very high with only 5% coupled product from the terephthaldehyde termination reaction contaminating the desired block copolymer product. Both blocks are formed by living processes so that overall polydispersity is low, with a typical figure of 1.2–1.3 being claimed. This process is therefore one of the few examples known to date of a quantitative transformation process.

It is perhaps germane to note that materials produced using this route have the desirable property of possessing hydrophobic and hydrophilic segments, so could have potential uses as dispersants, emulsifiers, non-ionic detergents, wetting agents and compatibilizing agents.

4.5.6 Radical to active monomer transformations

In the last few years Penczek has developed the concept of an active monomer polymerization mechanism, which can be used to generate narrow polydispersity polyoxiranes.[72,73] This mechanism, which can be described as a sequential addition of protonated oxirane to an alcohol, is shown in Scheme 4.7 (see also Section 1.3.3).

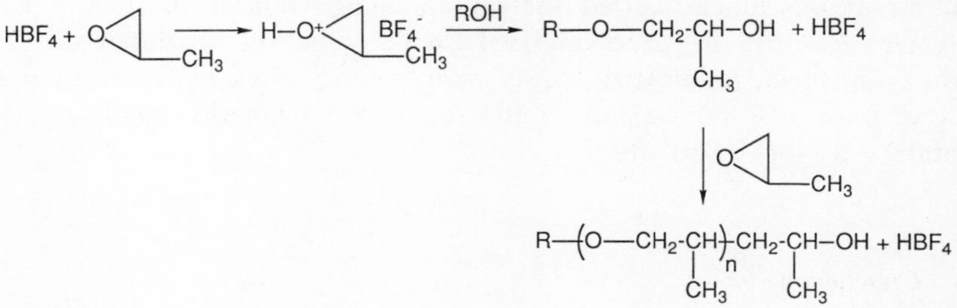

Scheme 4.7

By its very nature this process can be adapted so that polymeric alcohols are used as initiators. Thus Stewart[74] has shown that by using a free radically prepared hydroxyl terminated polybutadiene as a macroinitiator, block copolymers are quantitatively produced when trace amounts of a protonic acid are used in conjunction with epichlorohydrin (equation 4.41).

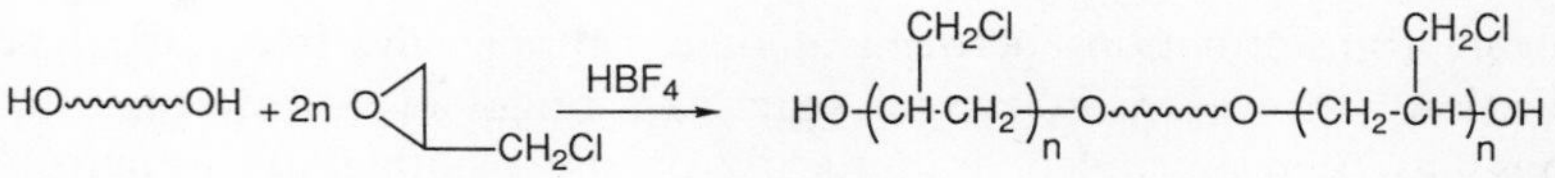

(4.41)

4.5.7 Transformation between radical polymerization and polypeptide synthesis

Over the last few years the synthesis of polymers showing biocompatibility has been of great interest. Among them, segmented polymers consisting of both hard and soft segments have attracted much attention as antithrombogenic materials. Bamford *et al.* have exploited the synthetic versatility of transformation reactions in order to produce a series of such materials.

Two complementary routes have been adopted which allow for transformations in both directions between radical polymerization and polypeptide synthesis. The first reported route involved the halo-acetylation of the terminal amino group on a poly(amino acid) such as sarcosine (Sar) in order to generate a halofunctionalized polypeptide.[75] This was then reacted, in the manner described for anion to radical transformation in Section 4.4.1 (equation 4.26), with an irradiated metallo-carbonyl species such as dimanganese decacarbonyl, to produce a radical macroinitiator. This then copolymerizes with styrene to give high, but not quantitative, conversion to the ABA block copolymer.

The method devised to undertake the reverse transformation, that is from radical to polypeptide, is equally elegant. Building on work by Gallot *et al.*,[76-78] who used anionic techniques to generate an amine functionalized polystyrene, Bamford and Tanaka produced an amine terminated polyvinylic compound such as poly(methyl methacrylate) by use of radical polymerization and an amino functionalized mercaptan chain transfer reagent.[79] This polymer was then copolymerized with γ-benzyl L-glutamate N-carboxyanhydride (Glu(OBzl) NCA), using a nucleophilic addition reaction, to give over 90% conversion to the requisite poly(methyl methacrylate-peptide) AB block copolymer.

4.6 Conclusions

A large number of transformation reactions has been investigated by various workers over the last few years. The general approach, proposed by Richards, involving a three-stage transformation process, permits a wide variety of polymerization mechanisms to be combined. This three-stage process, involving chain functionalization of block A, its isolation, then its reaction to give a macroinitiator which copolymerizes block B, is a powerful process and may be achieved using functionalization by initiation, termination or transfer. Most work has concentrated on functionalization by termination in systems involving transformations between anionic, cationic and free radical polymerizations. However, in recent years the range of transformation reactions has been expanded dramatically by use of other functionalization techniques and polymerization procedures.

Although no material produced by a transformation process is yet in commercial use, several routes allow access to previously unobtainable but desirable copolymers. It will therefore only be a matter of time before such technology becomes commercial. In many cases the transformation reaction has been demonstrated using only one or two common monomers. It is a great strength of this technology that by application of forethought, process modifications, or development of different conditions, this work can be expanded to include almost all possible monomers and their combinations. Thus the range of possible linear block copolymers accessible through these routes is vast, and ever increasing as new transformation strategies are developed in this fast growing field.

The work described in this chapter is not intended to define the limitations of transformation technology, but rather to indicate what may be achieved by adopting this methodology. Indeed, it is deliberately intended to spur further research in this potentially vast field, so that other new transformations, or improvements to existing ones, are investigated. Of particular interest must be transformations involving the newly discovered carbocationic, and group transfer living polymer systems, since these routes allow access to potentially ideal block copolymers.

To conclude, therefore, this chapter has indicated routes for the synthesis of a wide range of linear block copolymers in high yield. It is, however, early days for this technology. As awareness of these procedures grows, a vast increase in the use of transformation reactions will take place. New methodologies and products will result, as new homopolymerization mechanisms are harnessed for use in transformation reactions.

References

1. Reiss, C., Hurtrez, G. and Badahur, P. Block Copolymers, in *Encyclopedia of Polymer Science and Engineering*, J. Wiley & Sons, New York, Vol. 2, p. 324 (1985).
2. Goodman, I. *Comprehensive Polymer Science* (Eds. Allen, G. and Bevington, J.C.) Pergamon Press, Oxford, Vol. 6, p. 369 (1989).
3. Beyer, O., Müller, E., Peterson, S., Piepenbrink, H.F. and Windemuth, E. *Angew. Chem.* **62**, 57 (1950).
4. Vaughn, T.H., Suter, H.R., Lundsted, L.G. and Kramer, M.G. *J. Amer. Oil Chem. Soc.* **28**, 294 (1951).
5. Szwarc, M. *Nature* **178**, 1168 (1956).
6. Szwarc, M., Levy, M. and Milkovich, R. *J. Amer. Chem. Soc.* **78**, 2656 (1956).
7. Szwarc, M. *Carbanions, Living Polymers and Electron Transfer Process*, Wiley Interscience, New York (1968).
8. Szwarc, M. *Adv. Polym. Sci.* **49**, 1 (1983).
9. Young, R.N., Quirk, R.P. and Fetters, L.J. *Adv. Polym. Sci.*, **56**, 1 (1984).
10. Fontanille, M. in *Comprehensive Polymer Science* (Eds. Allen, G. and Bevington, J.C.) Pergamon Press, Oxford, Vol. 3, pp. 365 and 425 (1989).
11. Richards, D.H. and Szwarc, M. *Trans. Faraday Soc.* **55**, 1644 (1959).
12. Inoue, S. and Aida, T. in *Ring Opening Polymerization*, Elsevier, New York, Vol. 1, p. 185 (1984).
13. Miyamoto, M., Sawamoto, M. and Higashimura, T. *Macromolecules* **17**, 265 (1984).

14. Higashimura, T. and Sawamoto, M. *Adv. Polym. Sci.* **62**, 49 (1984).
15. Higashimura, T. and Sawamoto, M. in *Cationic Polymerization and Related Processes*, Academic Press, London, pp. 77 and 89 (1984).
16. Miyamoto, M., Sawamoto, M. and Higashimura, T. *Macromolecules* **17**, 2228 (1984).
17. Faust, R. and Kennedy J.P. *Polym. Bull.* **15**, 317 (1986).
18. Kennedy, J.P. *Macromolecular Preprints from Macromolecules* **86**, 8 (1986).
19. Burgess, F.J.,Cunliffe, A.V.,Richards, D.H. and Sherrington, D.C. *J. Polym. Sci., Polym. Lett. Ed.* **14**, 471 (1976).
20. Richards, D.H., *Brit. Polym. J.* **12**, 89 (1980).
21. Richards, D.H., Kingston, S.B. and Souel, T. *Polymer* **19**, 68 (1978).
22. Richards, D.H., Kingston, S.B. and Souel, T. *Polymer* **19**, 806 (1978).
23. Abadie, M.J.M., Burgess, F.J., Cunliffe A.V. and Richards, D.H. *J. Polym. Sci., Polym. Lett. Ed.* **14**, 477 (1976).
24. Cohen, P., Abadie, M.J.M., Schué, F. and Richards, D.H., *Polymer* **22**, 1316 (1981).
25. Abadie, M.J.M., Schué, F., Souel, T., Hartley, D.B. and Richards, D.H. *Polymer* **23**, 445, (1982).
26. Richards, D.H. *ACS Symp. Ser.* **286**, 87 (1985).
27. Burgess, F.J. and Richards, D.H. *Polymer* **17**, 1020 (1976).
28. Burgess, F.J., Cunliffe, A.V., MacCallum, J.R. and Richards, D.H. *Polymer* **18**, 719 (1977).
29. Burgess, F.J., Cunliffe, A.V., MacCallum, J.R. and Richards, D.H. *Polymer* **18**, 726 (1977).
30. Burgess, F.J., Cunliffe, A.V., MacCallum, J.R. and Richards, D.H. *Polymer* **18**, 733 (1977).
31. Franta, E., Lehmann, J., Reibel, L.C. and Penczek, S. *J. Polym. Sci.* **56**, 139 (1976).
32. Cohen, P., Abadie, M.J.M., Schué, F. and Richards, D.H. *Polymer* **23**, 1105 (1982).
33. Cohen, P., Abadie, M.J.M., Schué, F. and Richards, D.H. *Polymer* **23**, 1350 (1982).
34. Richards, D.H., Kingston, S.B. and Souel, T. *Polymer* **19**, 68 (1978).
35. Richards, D.H., Kingston, S.B. and Souel, T. *Polymer* **19**, 806 (1978).
36. Kucera, M., Salijka, Z. and Hudec, P. *Polymer* **26**, 1733 (1985).
37. Reeb, R., Vinchon, Y., Riess, G., Catula, J. and Brossas, J. *Bull. Soc. Chim. France* **11-12**, 2717 (1975).
38. Catula, J.M., Riess, G. and Brossas, J. *Makromol. Chem.* **178**, 1249 (1977).
39. Sumitomo Chemical Co., Osaka, Japan, Ger. Offen., 200.90.66 (1971).
40. Vinchon, Y., Reeb, R. and Riess, G. *Eur. Polym. J.* **12**, 317 (1976).
41. Riess, G. and Reeb, R. *Polym. Prepr.* **21**, 55 (1980).
42. Nicolova-Nankova, Z., Palacin, F., Raviola, F. and Riess, G. *Eur. Polym. J.* **11**, 301 (1975).
43. Abadie, M.J.M. and Ourahmoune, D. *Brit. Polym. J.* **19**, 247 (1987).
44. Burgess, F. M.Sc. Thesis, University of St. Andrews (1976).
45. Abadie, M.J.M., Schué, F., Souel, T. and Richards, D.H. *Polymer* **22**, 1076 (1981).
46. Souel, T., Schué, F., Abadie, M.J.M. and Richards, D.H. *Polymer* **14**, 1292 (1977).
47. Cunliffe, A.V., Hayes, G.F. and Richards, D.H. *J. Polym. Sci. Polym. Lett. Ed.* **14**, 483 (1976).
48. Lindsell, W.E., Service, D.M., Soutar, I. and Richards, D.H. *Brit. Polym. J.* **19**, 255 (1987).
49. Bamford, C.H., Eastmond, G.C., Woo, J. and Richards, D.H. *Polymer* **23**, 643 (1982).
50. Eastmond, G.C., Parr, K.J. and Woo, J. *Polymer* **29**, 950 (1988).
51. Niwa, M., Higashi, N. and Okamoto, M. *J. Macromol. Sci* **A25**, 445 (1988).
52. Niwa, M., Higashi, N. and Okamoto, M. *J. Macromol. Sci* **A25**, 1515 (1988).
53. Yagci, Y. *Polym. Commun.* **26**, 7 (1985).
54. Akar, A., Aydogan, A.C., Talinli, N. and Yagci, Y. *Polym. Bull.*, **15**, 293 (1986).
55. Hizal, G. and Yagci, Y. *Polymer* **30**, 722 (1989).
56. Richards, D.H. and Stewart, M.J. Unpublished Results.
57. Eastmond, G.C. and Woo, J. *Proc. 28th I.U.P.A.C. Macromol. Symp.*, 194 (1982).
58. Bamford, C.H., Middleton, I.P., Al-Lamee, K.G. and Paprotny, J. *Brit. Polym. J.* **19**, 269 (1987).
59. Tait, P.J. in *Comprehensive Polymer Science* (Eds. Allen, G. and Bevington, J.C.), Pergamon Press, Oxford, Vol. 4, p. 1 (1989).
60. Aldissi, M. *J. Chem. Soc., Chem. Commun.* **20**, 1347 (1984).
61. Aldissi, M. and Bishop, A.R. *Polymer* **26**, 622 (1985).
62. Aldissi, M. *Synthetic Metals* **13**, 87 (1986).
63. Stowell, J.A., Amass, A.J., Beevers, M.S. and Farren, T.R. *Polymer* **30**, 195 (1989).
64. Aldissi, M., Hou, M. and Farrell, J. *Synthetic Metals* **17**, 229 (1987).

65. Amass, A.J., Bas, S., Gregory, D. and Mathew, M.C. *Makromol. Chem.* **186**, 325 (1985).
66. Amass, A.J. and Gregory, D. *Brit. Polym. J.* **19**, 263 (1987).
67. Agouri, E., Favier, C., Laputte, R., Philardeau, Y. and Rideau, J. *Symposium on Block and Graft Copolymerisations: Preprints* GFP, Mulhouse, p. 55, 1972.
68. Eastmond, G.C. and Grigor, J. *Makromol Chem., Rapid Commun.* **7**, 375 (1986).
69. Gilliom, L. and Grubbs, R.H. *J. Amer. Chem. Soc.* **108**, 733 (1986).
70. Cannizzo, L.F. and Grubbs, R.H. *Macromolecules* **20**, 1488 (1987).
71. Risse, W. and Grubbs, R.H. *Macromolecules* **22**, 1558 (1989).
72. Penczek, S., Kubisa, P., Matyjaszewski, K. and Szymanski, R. *Pure & Applied Chem.* 140 (1984).
73. Wojtania, M., Kubisa, P. and Penczek, S. *Makromol. Chem., Macromol. Symp.* **6**, 201 (1986).
74. Stewart, M.J. Unpublished Results.
75. Imanishi, Y., Tanaka, M. and Bamford, C.H. *Int. J. Biol. Macromol.* **7**, 89 (1985).
76. Billot, J.-P., Douy, A. and Gallot, B. *Makromol. Chem.* **177**, 1889 (1976).
77. Perly, B., Douy, A. and Gallot, B. *Makromol. Chem.* **177**, 2569 (1976).
78. Billot, J.-P., Douy, A. and Gallot, B. *Makromol. Chem.* **178**, 164 (1977).
79. Tanaka, M., Mori, A., Imanishi, Y. and Bamford, C.H. *Int. J. Biol. Macromol.* **7**, 173 (1985).
80. Hurley, J., Richards, D.H. and Stewart, M.J. *Brit. Polym. J.* **18**, 181 (1986).
81. Yagci, Y. *Polym. Commun.* **27**, 21 (1986).

5 Chemical modification of preformed polymers

F.G. THORPE

5.1 Introduction

Synthetic polymers are prepared with the intention of producing materials with particular properties. The materials may have commercial applications such as the production of membranes, fibres, mouldings, adhesives, coatings and paint formulations, etc., or they may serve as chemical reagents or chromatographic supports. Biomedical applications are becoming increasingly important. The desired properties may depend, at least in part, on the presence of specific functional groups either in the polymer backbone, or as substituents in pendant groups linked to the main chain. This functionality may be incorporated in two ways; either a suitably substituted monomer is polymerized, or the substituent is introduced at a later stage by chemical modification of a preformed polymer. The first method leads to a fully substituted product of known overall structure, i.e. each monomer unit contains the functionality (and so the 'degree of substitution' is 100%) in a clearly defined position. The second method produces a substance which is rarely fully functionalized (although occasionally the degree of substitution exceeds 100%), and where the precise positions of the functional groups along the polymer chain (or even within the pendant groups) are not known. Usually the substituents are assumed to be randomly distributed, and in effect the product is a random copolymer. The ease of modification naturally depends on the chemical nature of the starting material, but also on its physical form, particularly on whether it is linear (i.e. soluble), or cross-linked (i.e. insoluble). For insoluble polymers chemical reaction might occur only on the outer surface.

It would seem, at least superficially, that functionalized reactive polymers might best be prepared from a suitably substituted monomer. However, this is not always convenient, and is sometimes impossible; the monomers may be very expensive or difficult to synthesize, or the desired substituent may inhibit polymerization. For example, the presence of nitro or phenolic groups in the benzene ring of styrene interferes with free-radical polymerization. In addition, if the functional groups of a modified polymer are themselves to be utilized in the product, perhaps as polymer-supported reagents, a large proportion may be 'wasted' if they are in an inaccessible part of an insoluble polymer bead. It is therefore frequently cheaper and easier to prepare a functionalized polymer by the second method, that is the chemical modification of a preformed polymer. It is the purpose of this review to describe some methods by which this might be achieved.

The reasons for carrying out such modifications include:

1. Changing the physical properties to improve the biocompatibility, fire-retardancy, adhesion, or ability to blend with other polymers etc. Sometimes changes may be required only on the outer surface in order to alter properties such as solvent repellency or friction.
2. Preparing polymer-supported reagents. It is sometimes advantageous to use reagents in polymeric form. Spent and unreacted reagents are easier to separate from the desired products, and reagent recovery and recycling are facilitated; this is particularly important for expensive reagents and catalysts. A wide range of such substances has been prepared,[1-5] and some are commercially available.
3. Controlled release drugs and pesticides. This can be achieved by attaching the pharmacologically active unit to a polymer.[6]
4. Mechanistic studies. There are some reactions where the linking of the reacting groups to polymers facilitates the mechanistic interpretation.[7]

In this chapter space limitations do not permit detailed descriptions of all the applications of the materials produced. A number of books and reviews are already available covering specialist topics.[8-14]

5.2 Chemical reactions of polymers: general aspects

The term modification normally implies the inter-conversion or replacement of functional groups or atoms (including hydrogen) already present, although it may also be an isomerization of stereoisomeric configurations, such as *cis-trans* isomers or tacticity. The changes may be directly on the main-chain (backbone), or on pendant groups linked to the main chain, or specifically at the ends of the chain.

When conducting reactions on polymeric systems it is often assumed that macromolecules behave chemically in much the same way as their low molecular weight counterparts. While this is generally true, it must be remembered that the physical form of the bulk material, which depends to a large extent on the polymerization conditions, may be crucial in determining the extent of reaction, and the physical form may change during the reaction as a result of solvent penetration or cross-linking. Solvent penetration may not be uniform throughout the reaction; reacted parts of a macromolecular system may become more compatible with the medium and more accessible to dissolved reagents,[15-17] resulting in uneven reactivity. Side reactions can also be troublesome, particularly if they cause cross-linking of a linear polymer, as can happen in chloromethylation of polystyrene. Since substituents are often difficult to remove once they have been incorporated into a polymer, wherever possible modifications are carried out using the minimum number of reaction steps in order to prevent the build-up of substituents introduced by side-reactions. In addition there may be 'proximity' or 'neighbouring group' effects. Such effects are well known for low molecular weight species, and the same effects might also be observed in polymers, as for example the rearrangement of polychloromethylthiirane[18,19] (Scheme 5.1), where a chlorine originally in a pendant group becomes linked to the main chain.

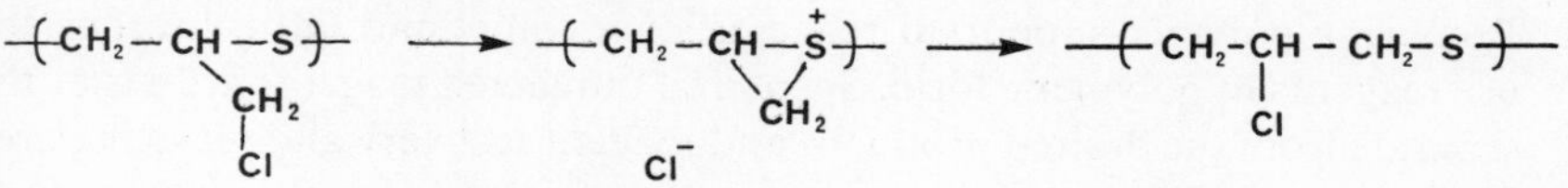

$$\mathrm{-\!\!\left(CH_2-CH-S\right)\!\!-} \quad \longrightarrow \quad \mathrm{-\!\!\left(CH_2-CH-\overset{+}{S}\right)\!\!-} \quad \longrightarrow \quad \mathrm{-\!\!\left(CH_2-CH-CH_2-S\right)\!\!-}$$

with pendant CH_2Cl / CH_2 Cl^- / Cl

Scheme 5.1

However, some effects appear to operate only in polymer systems. For example, the resistance of poly(methyl methacrylate) to complete hydrolysis is probably associated with the electronic or steric effects of the neighbouring groups[20] (although pure steric effects resulting from the presence of the polymer chain are likely to be important only when the functional groups are close to the polymer backbone).

When carrying out reactions on polymers it is not always easy to decide whether or not the modification has been successful. This is particularly true for cross-linked, insoluble resins, since many forms of spectroscopic analysis, such as NMR, are easily applicable only in solution (although satisfactory ^{13}C NMR spectra can be obtained for highly swollen lightly cross-linked polystyrenes[21] and other polymers[22]). Unreacted reagent may be occluded within the polymer matrix, and so elemental analyses need to be treated with caution. This difficulty is encountered, for example, in the mercuration of polystyrene and related materials. Even when it can be demonstrated that the new functional group is directly linked to the macromolecule, the precise positions and distributions are frequently unknown. A further difficulty is that substitution reactions are often incomplete, and the presence of unreplaced groups may complicate the analysis.

In general, the methods used to modify a polymer are those employed in the chemical conversions of low molecular weight species. This fact, taken together with the enormous range of synthetic polymers now available, with many different chemical structures, means that the subject of this review is exceptionally wide.

5.3 Modification of polymers

5.3.1 *Modifications on the polymer main chain*

Although all polymers can, in principle, undergo reaction on the main chain, this type of behaviour is most obvious for macromolecules such as poly-

CHEMICAL MODIFICATION OF PREFORMED POLYMERS   141

ethylene or polybutadiene which lack pendant groups. It is also important, however, for other polymers, such as aryl polysulphones and polyethers where substitution can occur in the aromatic rings which form an integral part of the backbone. The reactions of poly(vinyl halide)s also involve main chain modifications.

Some of the more commonly encountered systems are described in this section. Although the emphasis is naturally on processes aimed at effecting specific chemical changes, it must be remembered that some reactions, including those resulting in modification of the main chain, may occur as unwanted side reactions. These may take place during reactions on pendant groups, as in metallation of the phenyl groups of polystyrene.

5.3.1.1 *Polyalkenes and polydienes* For the simpler polymers such as polyethylene, modification is necessarily on the backbone. The direct reaction with halogens to give halogenated polyethylenes[23] is a well known procedure. The reaction can be of interest in studies of polymer blends, as chlorination sometimes improves miscibility of polyethylene with other polymers. Polymer supported transition-metal catalysts have also been prepared by reacting brominated polyethylene with lithium diphenylphosphide, and coordinating the ligand thus formed with palladium salts.[24]

Radiation grafting onto polyethylene is also a well known technique. The polyalkene and a monomer, usually in the presence of solvent, are simultaneously irradiated, usually using a ^{60}Co source.[25] Styrene and acrylate systems have been grafted onto the polyethylene backbone in this way. The technique has also been used to graft polystyrene onto the surface of polytetrafluoroethylene beads to produce supports used in oligonucleotide synthesis.[26]

The functional group already present in polybutadiene leads to a more varied chemistry. Although additions are typical, substitutions (of hydrogen) can also occur. For example, although bromination of poly(isobutylene-co-(2,3-dimethyl-1,3-butadiene)) gives predominant addition across the double bond, chlorination gives substitution, with some rearrangement of the double

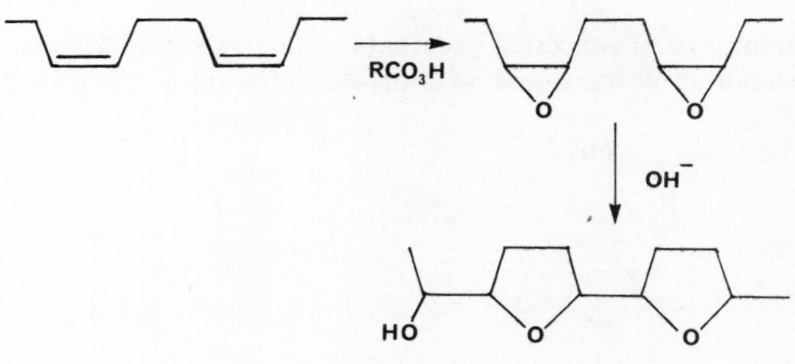

Scheme 5.2

bond.[27] Normally the additions are typical of simple alkenes, but the formation of certain groups in close proximity along the chain can sometimes result in, or at least facilitate, further reaction. An example of this is the formation of tetrahydrofuran structures when the product of peroxyacid oxidation of polybutadiene is treated with base[28] (Scheme 5.2).

Electrophilic, free-radical and concerted additions have all been studied. An interesting example of electrophilic addition is nitromercuration using mercuric chloride and sodium nitrite.[29] Between 20 and 30% of the alkene units undergo addition (Scheme 5.3).

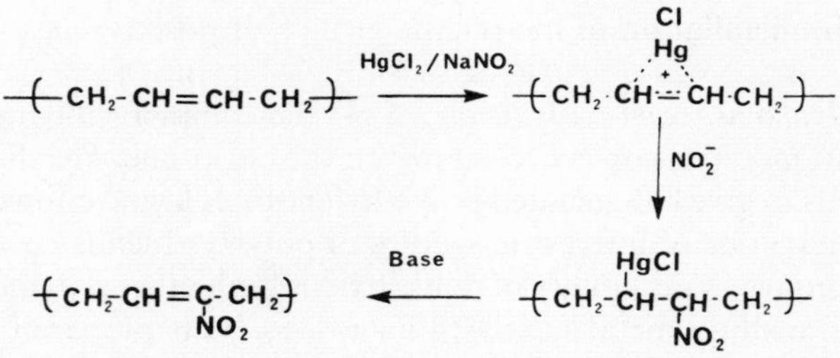

Scheme 5.3

Later demercuriation under basic conditions can lead to regeneration of the double bond, but with retention of the nitro group the overall reaction is replacement of an olefinic hydrogen by a nitro group. A wide range of concerted additions has been examined. Hydroboration, for example, followed by cyanomethylation, can be used to introduce a functionalized pendant group[30] (Scheme 5.4).

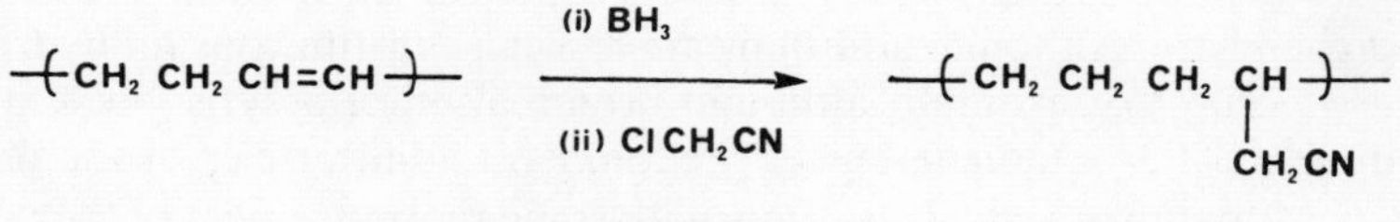

Scheme 5.4

A different type of concerted process (an 'ene' reaction) has been observed in the reaction of polyisoprene with maleic anhydride[31] (Scheme 5.5).

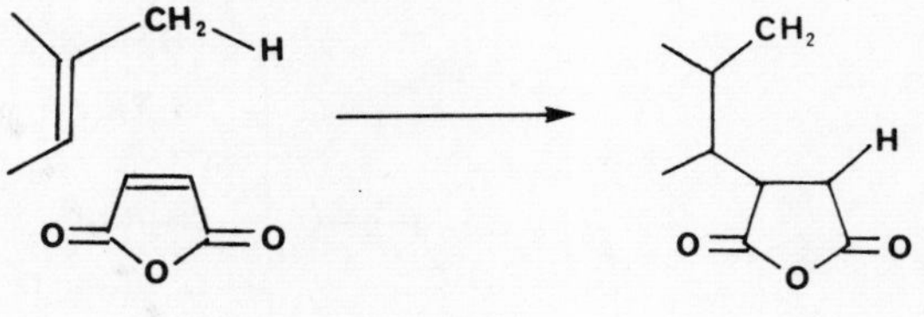

Scheme 5.5

CHEMICAL MODIFICATION OF PREFORMED POLYMERS 143

A third concerted process which has been studied is the reaction with carbenes to produce polymers containing cyclopropyl systems.[32] Hydroformylation of the double bond in polyenes can also be carried out quite cleanly using rhodium complexes as catalysts.[33]

5.3.1.2 *Poly(vinyl chloride)* Poly(vinyl chloride) can undergo substitution, elimination, and metallation. Elimination reactions can, in fact, cause difficulties in manufacture, since dehydrochlorination to give a polymer containing an alkene function can occur during processing, unless steps are taken to avoid this. A second HCl is favourably eliminated adjacent to the initially formed double bond, and polyenes are produced.[34,35]

Although substitution usually implies replacement of the chlorine atom, replacement of hydrogen is also possible. For example, in some recent studies on polymer miscibilities, poly(vinyl chloride) was chlorinated using free-radical conditions (moderate chlorination was found to improve the miscibility with poly(methyl methacrylate)[36]). Nucleophilic displacement of chlorine, however, is not easily achieved unless aprotic solvents such as dimethylformamide are used.[37,38] In such solvents nucleophiles such as thiolates, xanthates, dithiocarbamates and azides have been used with success (Scheme 5.6). Some of these substitutions probably involve neighbouring group effects.[37] Similar reactions have been reported[39] for polychlorotrifluoroethylene, the chlorine being replaced by sulphur, selenium or phosphorus nucleophiles. Metallation of poly(vinyl chloride) by alkyl lithium compounds might at first appear to be a potentially useful route to a range of functionalized polymers. However, the reaction is relatively complex, and the chlorine may be replaced by either the alkyl (from the organolithium reagent) or the metal[40] (Scheme 5.7). The organolithium species (1) can undergo a Wurtz-type coupling with an unreacted chlorine atom to give a cross-linked polymer. Despite these difficulties functionalized materials have been prepared via lithiation of poly(vinyl chloride)[41] (Scheme 5.8), and surface modifications of polychlorotrifluoroethylene using methyllithium have been

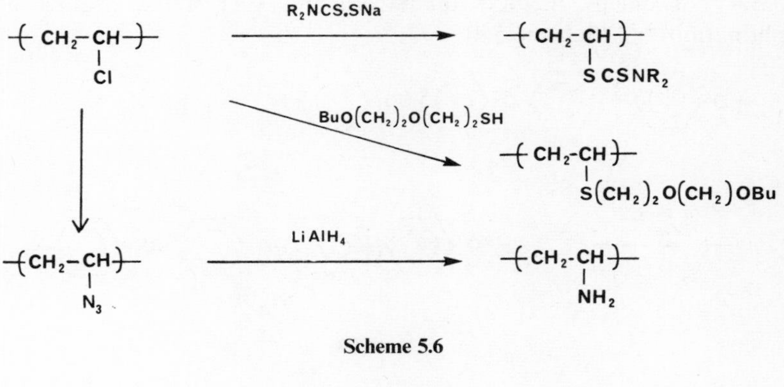

Scheme 5.6

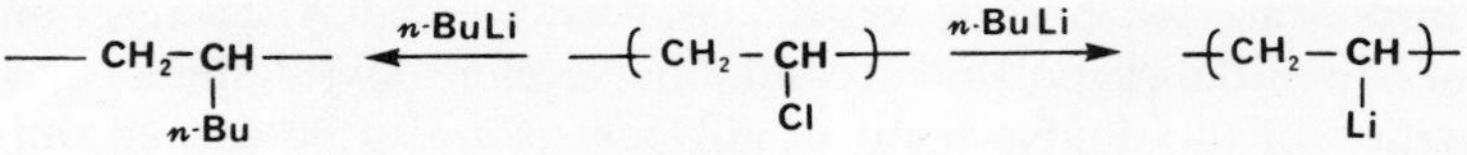

144   NEW METHODS OF POLYMER SYNTHESIS

$$\text{---CH}_2\text{-CH---} \xleftarrow{n\text{-BuLi}} \text{--(CH}_2\text{-CH)--} \xrightarrow{n\text{-BuLi}} \text{--(CH}_2\text{-CH)--}$$
$$\quad\quad |\quad\quad\quad\quad\quad\quad\quad |\quad\quad\quad\quad\quad\quad\quad |$$
$$\quad\ n\text{-Bu}\quad\quad\quad\quad\quad\quad\text{Cl}\quad\quad\quad\quad\quad\quad\text{Li}$$

(1)

Scheme 5.7

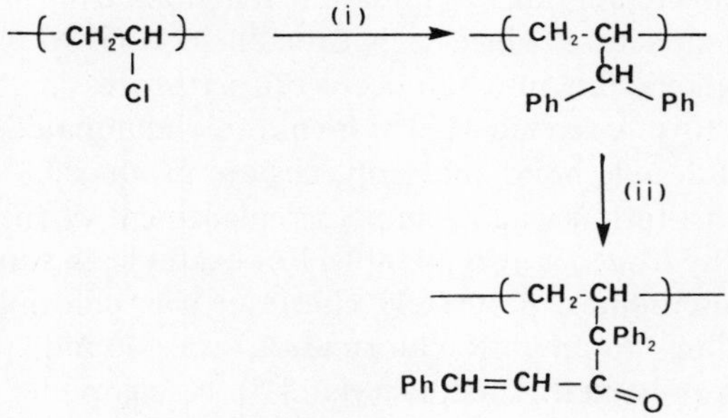

$$\text{--(CH}_2\text{-CH)--} \xrightarrow{(i)} \text{--(CH}_2\text{-CH)--}$$
$$\quad\quad |\quad\quad\quad\quad\quad\quad\quad\quad\quad\quad |$$
$$\quad\ \text{Cl}\quad\quad\quad\quad\quad\quad\quad\quad\text{CH}$$
$$\quad\quad\quad\quad\quad\quad\quad\quad\quad\text{Ph}\nearrow\quad\nwarrow\text{Ph}$$

$\downarrow$ (ii)

$$\text{---(CH}_2\text{-CH)---}$$
$$\quad\quad\quad\quad\quad |$$
$$\quad\quad\quad\quad\quad\text{CPh}_2$$
$$\text{PhCH=CH--C}\!\!\diagdown_{\!\!O}$$

(i) Ph$_2$CHLi   (ii) Na naphthalene then PhCH=CHCOCl

Scheme 5.8

reported. A complex sequence of reactions involving metal chlorine exchange, loss of lithium fluoride to give an alkene, and subsequent addition of methyllithium, results in incorporation of methyl groups and double bonds in the outer surface.[42]

5.3.1.3 *Polysulphones*   Poly(aryl ether sulphone)s, e.g. structure (2), are of great commercial importance. They form a family of thermoplastic materials which can be used for prolonged periods at relatively high temperatures[43] and they have also been used as composites or single component membranes for water desalination.[44] Ionic groups incorporated into the aromatic rings in the polymer chain significantly affect the physical properties of the polymers, particularly water absorption (which itself can affect the mechanical properties). A convenient method of introducing such ionic groups is by sulphonation[44,45] (Scheme 5.9).

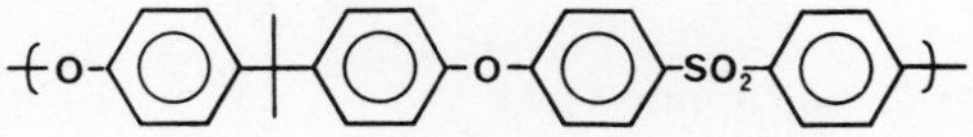

$$\text{--(O--}\bigcirc\text{--}\!\!\!|\!\!\!\text{--}\bigcirc\text{--O--}\bigcirc\text{--SO}_2\text{--}\bigcirc\text{--)--}$$

(2)

$$\xrightarrow{SO_3}\quad\text{--(O--}\bigcirc\text{--}\!\!\!|\!\!\!\text{--}\bigcirc\text{--O--}\bigcirc\text{--SO}_2\text{--}\bigcirc\text{--)--}$$
$$\quad\quad\quad\quad\quad\quad\quad\quad\quad\quad\quad\quad |$$
$$\quad\quad\quad\quad\quad\quad\quad\quad\quad\quad\text{SO}_2\text{OH}$$

Scheme 5.9

CHEMICAL MODIFICATION OF PREFORMED POLYMERS 145

The sodium salt is more stable than the free acid form. Chlorosulphonic acid can be used as the sulphonating agent, but better results are obtained with a sulphur trioxide/triethyl phosphate complex.

Other modifications of the benzene ring are possible, using reactions which will be described later for polystyrene. For example, the chloromethyl group can be introduced, further modification then leading to poly(ether sulphone)s containing pendant vinyl groups[46] (Scheme 5.10).

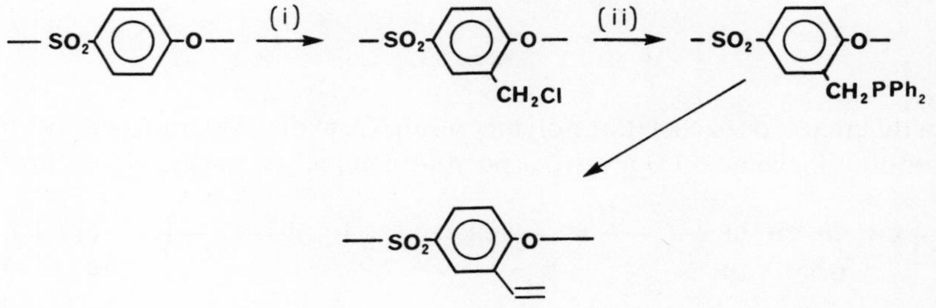

(i) $ClCH_2OCH_3/SnCl_4$ (ii) PPh_3

Scheme 5.10

Lithiation is possible in some aliphatic systems. Lithiation of poly(alkene sulphone)s, followed by alkylation, has been studied because of the potential use of these materials in high resolution lithography[47] (Scheme 5.11).

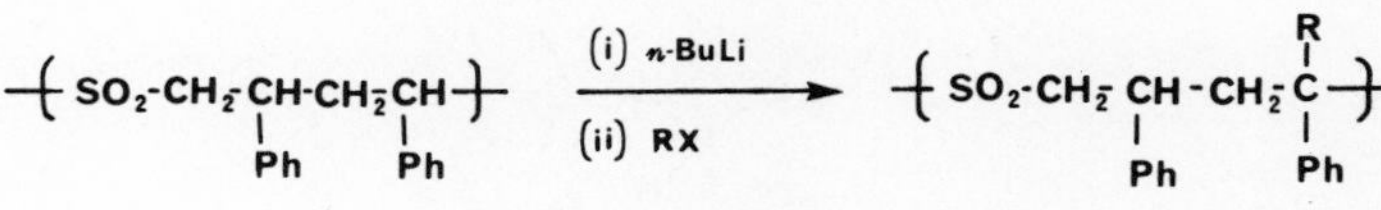

Scheme 5.11

5.3.2 Modification of pendant groups

Many synthetic and natural polymers contain pendant groups such as hydroxyls, carboxylic acid or ester functions, or aryl groups attached to the main chain. Numerous methods are available for producing chemical changes in these macromolecules; the entire pendant group may be replaced (when the process is a modification of the backbone), but frequently the changes involve addition to or substitution within the pendant unit. The more common procedures will be described briefly.

5.3.2.1 Poly(vinyl alcohol)
Certain reactions of poly(vinyl alcohol) involve direct changes on the main chain, and so might properly be considered above. However, for convenience they are described in this section. Such reactions

include the thermal and acid catalysed dehydrations to produce a polyene[48] and substitution where the entire hydroxyl group is replaced. For example, treatment with thiourea (Scheme 5.12) and base gives a polymer containing mercaptan groups[49,50] which can be used in the removal of mercury(II) and copper(II) ions from aqueous solution.

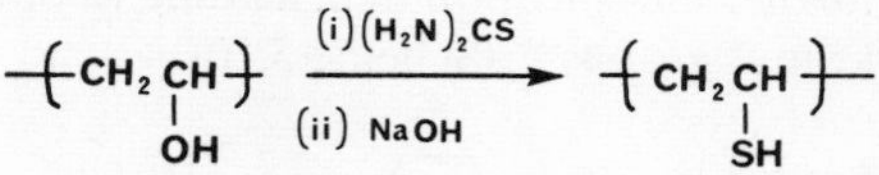

$$\{CH_2CH(OH)\} \xrightarrow[\text{(ii) NaOH}]{\text{(i) }(H_2N)_2CS} \{CH_2CH(SH)\}$$

Scheme 5.12

A different type of chelating polymer results from the oxidation of poly(vinyl alcohol)[51] (Scheme 5.13) to give a poly(β-diketone) or poly(enol-ketone) (3).

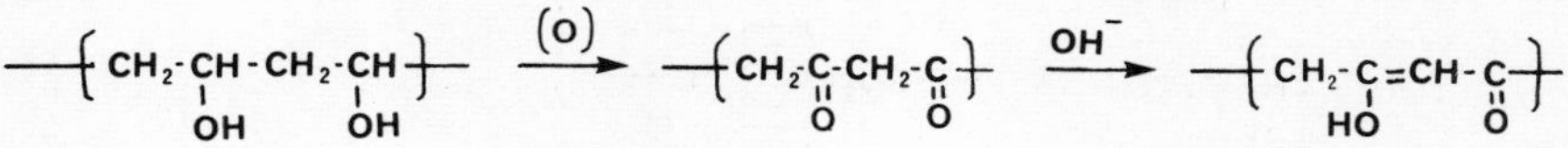

$$\{CH_2\text{-}CH(OH)\text{-}CH_2\text{-}CH(OH)\} \xrightarrow{(O)} \{CH_2\text{-}C(O)\text{-}CH_2\text{-}C(O)\} \xrightarrow{OH^-} \{CH_2\text{-}C(HO)=CH\text{-}C(O)\}$$
(3)

Scheme 5.13

The product, (3), complexes with various transition metal ions, such as nickel. However, in order to form a complexing polymer, it is not always necessary to introduce new ligand systems; direct reaction of unmodified poly(vinyl alcohol) can also occur. Examples include coordination to copper(II)[52] and reactions with boric acid and borate salts.[53-55] Often, as a result of these reactions, cross-linking of the polymer takes place (see (4)).

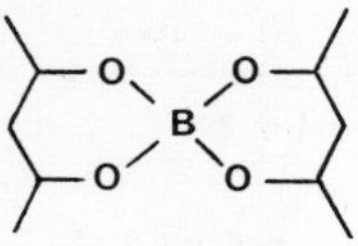

(4)

A considerable amount of work has been carried out on the preparation of biologically active forms of poly(vinyl alcohol) by attaching, for example, pesticides or herbicides to the polymer via the alcohol function. Controlled release of the active component then occurs in an aqueous environment.[56,57] The most common method of attachment is by forming an ester with an acid or acid chloride form of the active substance (Scheme 5.14), but use can be made also of the ability of poly(vinyl alcohol) to form acetals. Sulphonamide groups have been introduced in this way (Scheme 5.15). The acetal-forming reaction of poly(vinyl alcohol) has been used commercially for many years in the production of poly(vinyl butyral) for the manufacture of safety glass.[57] During acetal formation, not all the hydroxyl groups react; some become isolated along the polymer chain, and the absence of a suitably placed second hydroxyl group prevents reaction with the aldehyde.

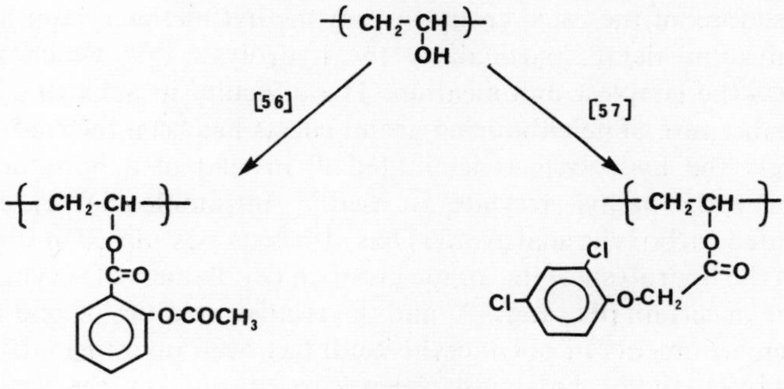

CHEMICAL MODIFICATION OF PREFORMED POLYMERS 147

Scheme 5.14

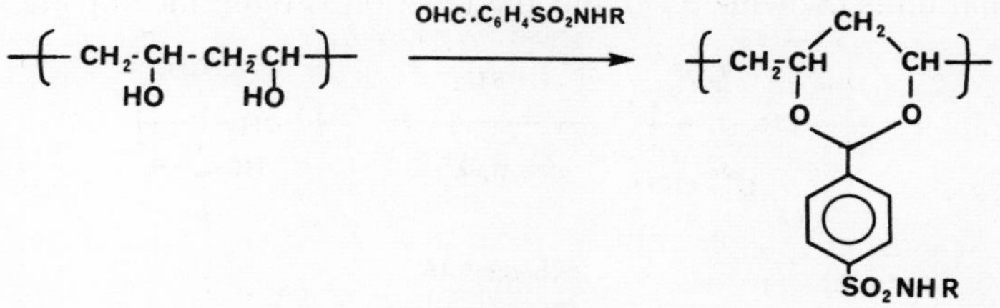

Scheme 5.15

Although the chemical modifications are usually carried out with the aim of producing a new material, the reactions are occasionally used for analytical purposes. For example, the surface concentration of hydroxyl groups in poly(vinyl alcohol) films has been determined by reacting with isocyanates to give bound carbamate functions, which liberate amines, identified spectroscopically, on hydrolysis.[58]

5.3.2.2 *Polyacrylates* This group of polymers includes the esters and amides of polymers derived from acrylic acid and methacrylic acid, as well as the acids themselves, and acrylonitrile. These polymers have many applications. Modifications can be carried out on the carbonyl or nitrile function, usually by nucleophilic addition, or by replacement on the alkoxy- or nitrogen-containing entities.

The polymers containing ester functions are usually methyl esters, but some are prepared from monomers already having additional functionality in the alcohol-derived part of the molecule, such as glycidyl methacrylate, (**5**), or hydroxyalkyl methacrylates, (**6**), i.e. the polymers are already 'bifunctional'.

The reactions of the ester group in poly(methyl methacrylate) have been studied in some detail, particularly the hydrolysis,[59,60] which might be regarded as the simplest modification. The difficulty in achieving complete hydrolysis because of neighbouring group effects has been referred to above; interestingly the hydrolysis is facilitated if, instead of a homopolymer, a copolymer with methyl acrylate is used.[61] Intramolecular catalysis (by undissociated carboxylic acid groups) has also been postulated in some recent studies on the hydrolysis of the amide group in copolymers of acrylic acid and acrylamide in certain pH ranges,[62] and the relative acidities of acid groups in different environments in poly(acrylic acid) has been investigated.[63]

Mechanistic studies have also been carried out on the reactions of organolithium compounds with poly(methyl methacrylate). The normal addition does occur to give (after hydrolysis) polymers containing alcoholic functional units (Scheme 5.16), but the reaction is often incomplete and the

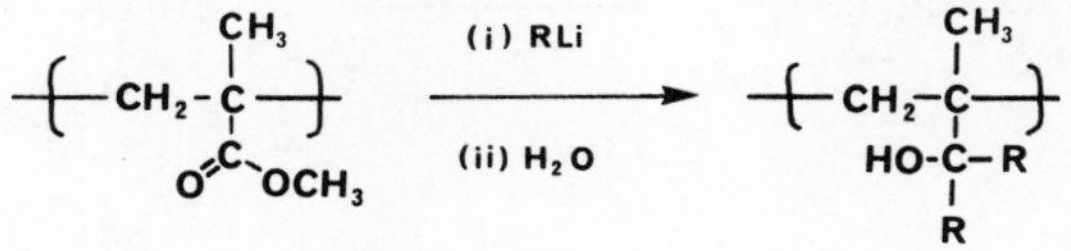

Scheme 5.16

products are complex and have poorly defined structures.[64] A different type of reaction has been reported if sulphur or heterocycle stabilized reagents of the type shown in Scheme 5.17 are used.[65,66] Some cyclization can also occur to

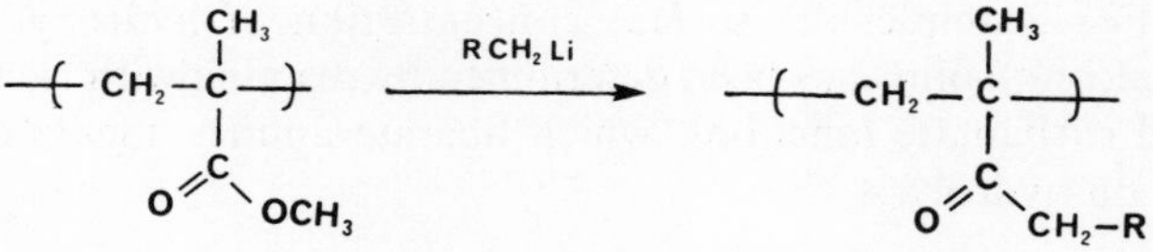

$R = C_6H_4N$, PhS or CH_3SO_2

Scheme 5.17

give polymers containing the structural unit (**7**). Similar cyclizations to give ladder polymers (**8**) are known[67] for polyacrylonitrile, although 'normal' addition of organolithium reagents to the nitrile group (to give (**9**) after hydrolysis) occurs in copolymers with isolated nitrile units.[66]

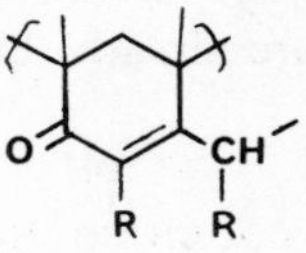

(7) 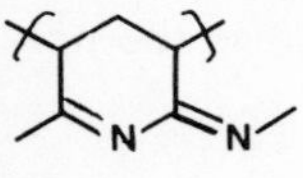   (8) 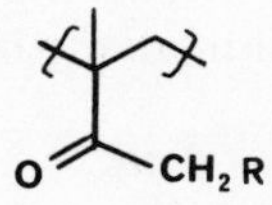   (9)

CHEMICAL MODIFICATION OF PREFORMED POLYMERS

Polyacrylates have a broad range of uses and, increasingly, pharmaceutical and medical applications in general are being recognized. For example, polymeric materials prepared using monomer (5) and related systems are useful as coatings and carriers for biologically active materials.[68] Scheme 5.18 shows how the hydroxyl group in macroreticular polymers prepared from 2-hydroxyethyl methacrylate can be utilized in standard procedures.[69] Simpler polymers, such as poly(methacryloyl chloride), which forms esters relatively easily, can also be used as carriers for biologically active substances[56] (Scheme 5.19).

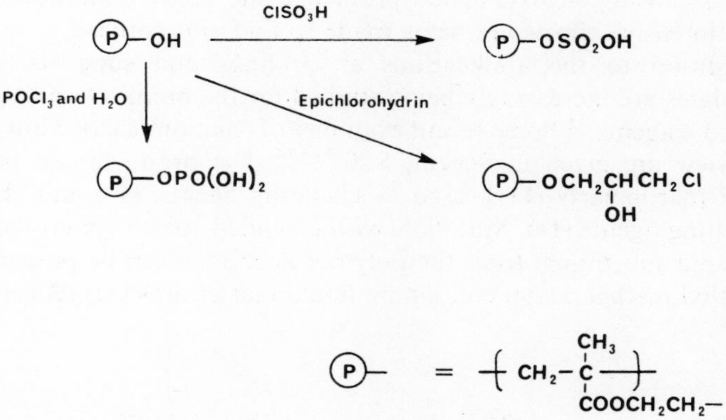

Scheme 5.18

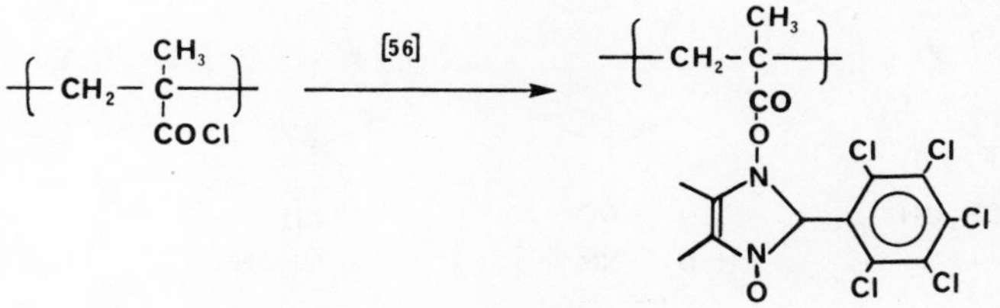

Scheme 5.19

Particularly well known polymers with biocidal activity, however, are the trialkyltin acrylates and methacrylates. These are used as marine anti-foulants and anti-slime agents,[56,70] and are sometimes prepared as copolymers with fluoroalkyl acrylates to produce oil- and water-repellant coatings. When used as marine paints, the tin-containing acrylate systems appear to have self-polishing properties, in addition to their biocidal functions. The materials may be manufactured from monomers containing the metal, but the tin can be incorporated into a preformed polymer by reacting the acid or acid chloride form with an organotin hydroxide. Cross-linking is avoided if a trialkyltin compound is used, giving a polymer containing structure (10).

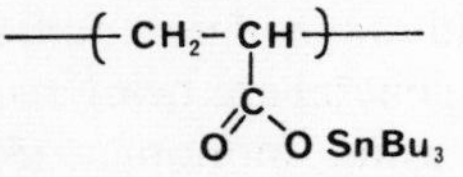

150 NEW METHODS OF POLYMER SYNTHESIS

$$-\!\!-\!\!(CH_2\!-\!CH)\!-\!\!-$$
$$\underset{O}{\overset{C}{|}}\;\;\underset{OSnBu_3}{}$$

(10)

The tin compounds are used in polymer-bound form in order to control and limit the amount of toxic material released into the marine environment. Release occurs via hydrolysis rather than by simple leaching of an additive from an insoluble polymer coating. Nevertheless, increased use of these antifoulants is causing environmental problems, and hence limitations on their use, and intensive efforts are being made to find alternatives.

In addition to the applications as coatings and supports, modified polyacrylates are increasingly being studied for the preparation of polymer-supported reagents.[71] Some recent examples of reactions carried out with this aim in view are given in Scheme 5.20.[71–74] The products can be further modified (particularly (11)), used as chelating agents, (12) and (13), or as halogenating agents (14). Materials with extended spacer groups separating the required substituent from the polymer backbone can be prepared using poly(methyl methacrylate) containing functional groups (11) (Scheme 5.21).

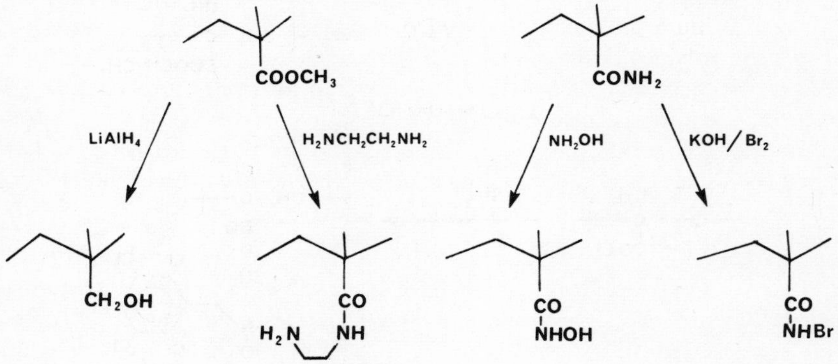

Scheme 5.20

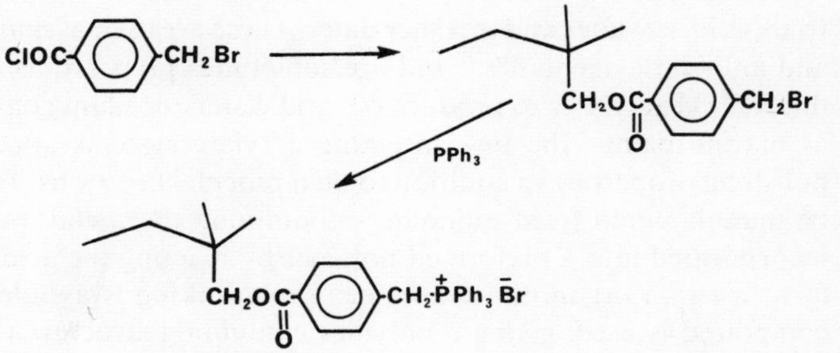

Scheme 5.21

5.3.2.3 *Polystyrene* Polystyrene is available in a number of different forms, both as linear and as cross-linked resins prepared by copolymerizing styrene with divinyl benzene. For the insoluble materials the pore size and surface area can be varied over a wide range.[18,75] The cross-linked polymers may be gel-type, with a low degree of cross-linking, or macroreticular structures, and the physical form can be of great importance in determining how easily chemical modifications may be carried out.[16] In addition to styrene itself, monomers containing substituents on the phenyl ring are available, and from some of these functionalized polystyrenes can be prepared. However, not all the substituted monomers can conveniently be polymerized (see above).

Polystyrene was one of the first reported synthetic polymers, and so, not surprisingly, it provides some of the earliest examples of the use of synthetic polymers as chemical reagents or chromatographic materials,[14] including acidic and basic ion-exchange resins. In 1963, Merrifield used modified polystyrenes in his classic work on polymer-supported peptide syntheses.[76]

At normal temperatures, and in many reaction conditions, polystyrene is thermally and chemically stable. However, the aromatic ring is not chemically inert, and can be modified under conditions which do not result in polymer degradation. Typically the ring undergoes classic electrophilic substitutions such as halogenations and Friedel–Craft reactions, and metallations to introduce lithium, mercury, thallium or tin-containing units. Usually, the substituted polymer produced in these initial reactions is modified further to produce materials with the functionality required for specific applications. For example, an initially introduced lithium atom might be replaced by a diphenylphosphine group to be used as a chelating ligand. It is beyond the scope of this chapter to describe these 'secondary modifications' in detail, and attention will be focused on the initial reactions.

Bromination. A convenient procedure for the bromination of both linear and cross-linked polystyrene is to use bromine and thallium(III) triacetate in tetrachloromethane.[77,78] Substitution is predominantly in the *para*-position. The degree of substitution obtained is *c.* 80%. The polymer thus formed is frequently used to prepare reagents containing diphenylphosphine residues[79] or lithiated polystyrene (see below).

Chloromethylation. Chloromethylation, followed by reaction with nucleophiles, is one of the most important methods for introducing functionality into polystyrene. The polymer is reacted with a chloromethyl ether and a Lewis acid such as tin(IV) chloride[80] (Scheme 5.22) to give products containing

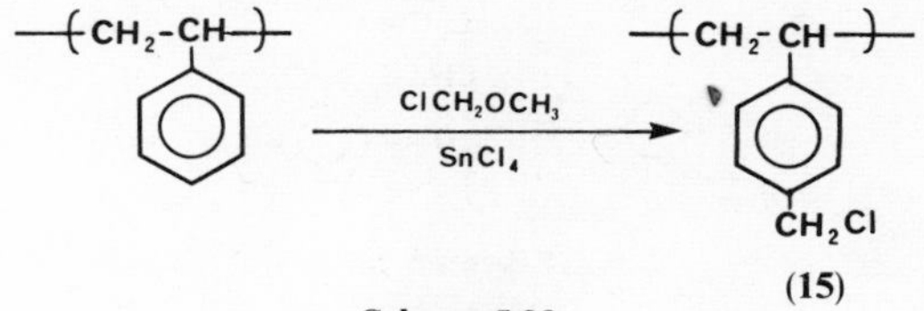

Scheme 5.22

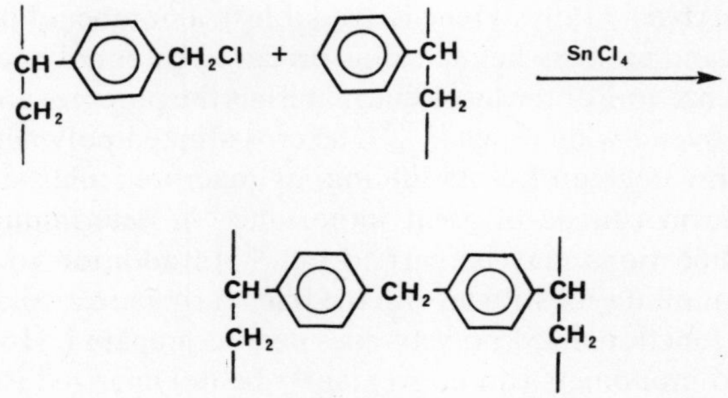

Scheme 5.23

structure (15), with a degree of substitution of c. 55% (less substitution is obtained if zinc chloride or boron trifluoride etherates are used as catalysts[81]). If a linear polymer is used problems arise with cross-linking, and the polymer may gel (Scheme 5.23). This reaction is of less consequence with cross-linked

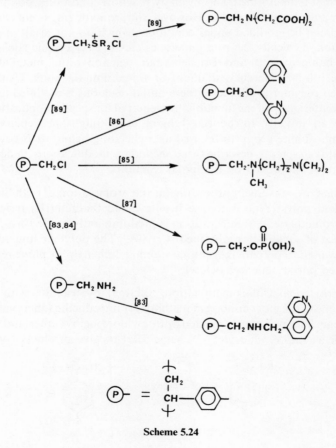

Scheme 5.24

CHEMICAL MODIFICATION OF PREFORMED POLYMERS 153

resins, although it undoubtedly occurs and produces a higher degree of cross-linking.

Standard procedures can be used to oxidize the chloromethyl group into aldehyde or carboxylic acid functions, or the chlorine atom can be replaced via nucleophilic substitution.[81-89] Some recent examples of these substitutions, showing the incorporation of more complex functional groups, are shown in Scheme 5.24. The modifications were usually carried out with the intention of preparing supported protecting groups[1,2] or coordinating ligands.[89]

Acylation. The Friedel–Craft reaction is used to introduce the acyl group into polystyrene. The group may contain only the carbonyl function[90] **(16)**, or it may be bifunctional as in **(17)**. This is of particular interest because it provides a method of attaching bifunctional or polyfunctional pendant groups,[91] including quite complex ligand systems (Scheme 5.25).

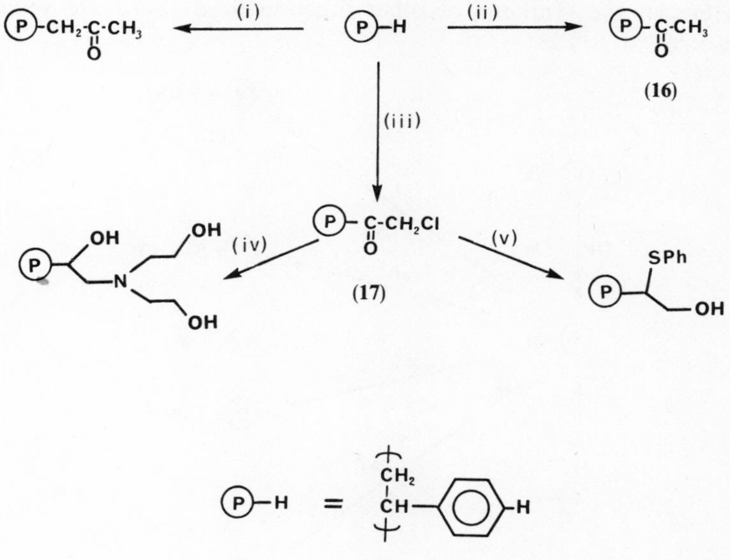

(i) BrCH$_2$COCH$_3$ (ii) CH$_3$COCl (iii) ClCOCH$_2$Cl
(iv) NaBH$_4$/NaH/HN(CH$_2$CH$_2$OH)$_2$ (v) NaBH$_4$/NaH/PhSH

Scheme 5.25

Metallation. Lithiation, like chloromethylation, provides another very versatile method of introducing a wide range of functional groups into polystyrene, but the initially formed material is further reacted with electrophiles rather than nucleophiles. Direct lithiation of cross-linked polymers using *n*-butyllithium is not successful, however, unless additives such as TMEDA are present, and an excess of the lithium reagent is used. Even then, only about

20% of the phenyl groups are metallated.[77,92–94] Higher degrees of metallation can be obtained if brominated polystyrene is metallated instead of polystyrene itself, and the site of metallation is more clearly defined. The brominated polymer needs to be treated several times with n-butyllithium, and the choice of swelling solvent is important. The best results are obtained using THF or benzene.[77,78,95]

Linear polystyrene can be lithiated directly using n-butyllithium in hydrocarbon solvents if TMEDA is present. However, the reaction is not without complications. In cyclohexane the reaction is described as slow and incomplete, and precipitation occurs after only a relatively low degree of lithiation has been obtained.[96,97] Nuclear magnetic resonance studies indicate that the metallation is almost entirely in the ring, with very little in the backbone,[98,99] unlike metallation of alkylbenzenes, in which appreciable side-chain lithiation is observed.[100] Examples showing the use of lithiated polystyrene in the synthesis of other functionalized resins are shown in Scheme 5.26.

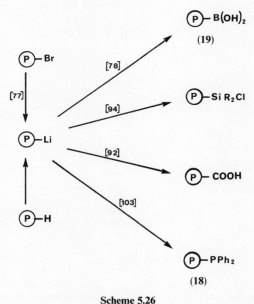

Scheme 5.26

Groups which have been successfully introduced include boronic acid,[78,101] carboxylic acid,[92] diphenylphosphine,[102,103] and silane.[94]

Two of the above reactions are worth describing in more detail. Resins containing the diphenylphosphino group (**18**) are important not only as ligands used in the preparation of polymer-supported metal catalysts[95,104,105] or organolithium cuprates,[106] but also as reagents in Wittig reactions[107,108] and for the conversion of carboxylic acids to acid chlorides[109] (Scheme 5.27).

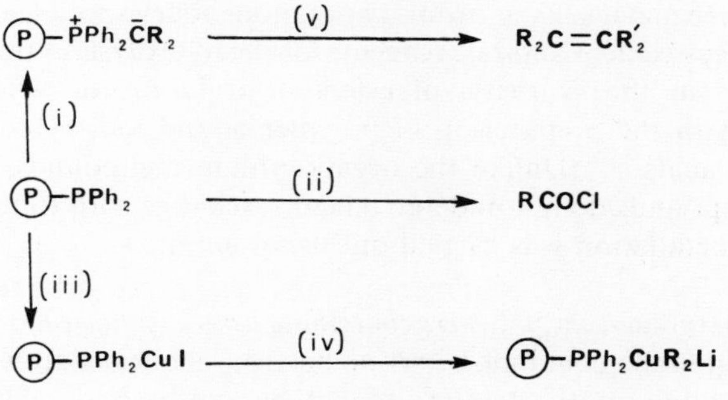

CHEMICAL MODIFICATION OF PREFORMED POLYMERS

[Scheme 5.27 diagram:]

(P)—⁺PPh₂C̄R₂ —(v)→ R₂C=CR′₂

↑ (i)

(P)—PPh₂ —(ii)→ RCOCl

↓ (iii)

(P)—PPh₂CuI —(iv)→ (P)—PPh₂CuR₂Li

(i) R₂CHX then base (ii) CCl₄/RCO₂H
(iii) CuI (iv) 2RLi (v) R′₂CO

Scheme 5.27

Resins containing the boronic acid groups (**19**) complex with diols and catechols to give (**20**), and have been used to remove selectively these materials from solution, or as protecting groups for the diol unit.[78,101]

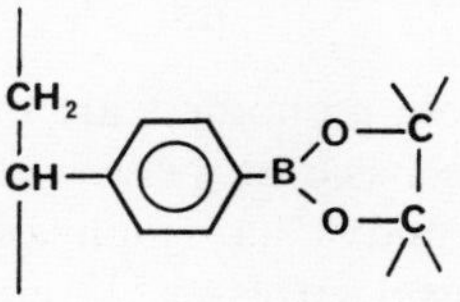

(**20**)

A different type of metallation of polystyrene is mercuriation or thalliation to give resins containing units (**21**) and (**22**).[110]

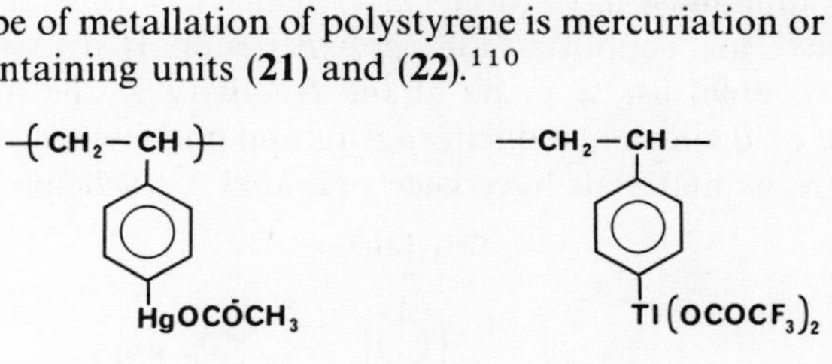

(**21**) HgOCOCH₃   (**22**) Tl(OCOCF₃)₂

Mercury(II) acetate gives low degrees of substitution for cross-linked (2%) polymers, and although the addition of perchloric acid catalyses the reaction and gives increased metallation, it also leads to the formation of mercury(I) salts, which are difficult to remove. The metal trifluoroacetates are more satisfactory, and 2% cross-linked polymers with degrees of substitution of 65% (for mercury) and 38% (for thallium) can be obtained without the need for a catalyst. The choice of solvent is again important and in the correct conditions controlled mercuriation limited to the polymer surface has been achieved.[111] Although these mercuriated polymers are expensive, the toxic risks are less than for the free metal salts or ordinary organomercury species, and they have been used as convenient and safe reagents for the hydrolysis of thioacetals and thioketals,[112] in the synthesis of esters of trifluoracetic acid[113] and as intermediates in the preparation of polymer-bound iodoxy compounds[114] and boronic acids.[110] (Unlike the organolithium compounds, the organomercury compounds do not undergo a facile exchange with trimethyl borate, so the transmetallation was carried out using borane.)

5.3.2.4 *Polystyrene-type polymers containing heterocyclic groups* A number of vinyl compounds containing aryl or heterocyclic rings are commercially available in addition to styrene, or can be synthesized relatively easily. Such materials include 4-vinylpyridine (23), 4-vinylimidazole (24), 2-vinylthiophene (25), and N-vinylcarbazole (26).

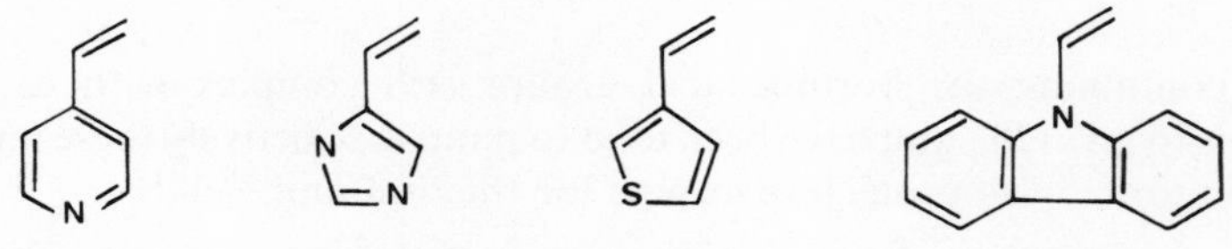

(23)   (24)   (25)   (26)

Polymers prepared using these monomers are frequently of interest in their own right. For example, the basic properties of polyvinylpyridine make it useful as an acid acceptor, and it has applications in the preparations of ion-exchange resins and polyelectrolytes,[115] and can be used as a ligand (for example, with copper salts[116]). However, some polymers prepared from the above compounds have been investigated as possible alternatives to polystyrene as supports for polymer-supported reagents. With poly(N-vinylpyridine) use is made of the reactivity at the nitrogen atom, and supported oxidizing and reducing agents and polymers of potential use in the design of resist materials have been prepared[117] (Scheme 5.28).

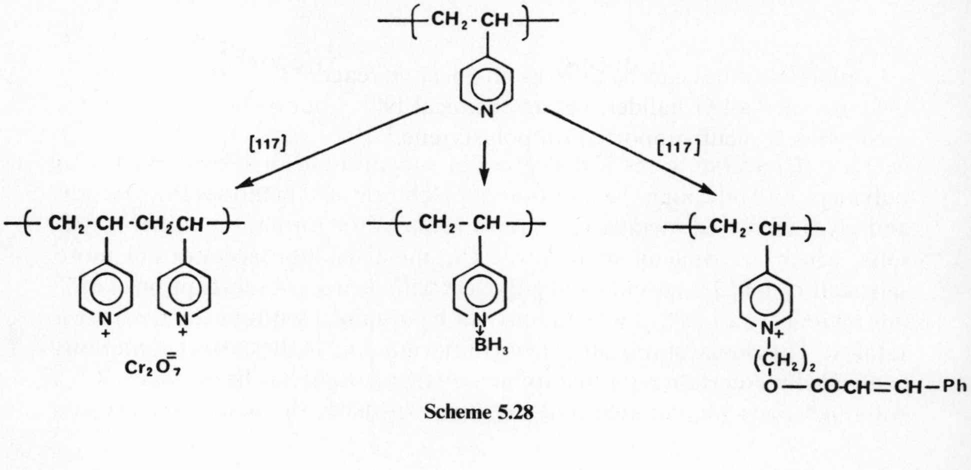

Scheme 5.28

Other polymers (such as those prepared from (**25**)) have been investigated because the reactivity of the aromatic ring might facilitate functionalization, as compared with polystyrene. On the other hand, the reagents formed may not all have the stability of the polystyrene-based materials.

The metallations of some of these heterocyclic compounds have been studied in some detail. Linear poly(N-vinylcarbazole) cannot be lithiated directly (using THF as the solvent), but the less direct route involving bromination followed by lithium-halide exchange (with *n*-butyllithium) gives a polymer with a degree of substitution of *c.* 50%, and mercuration and thalliation of the linear and cross-linked systems proceeds smoothly to give materials with higher degrees of substitution than for polystyrene.[118] Direct lithiation is successful for poly(2-vinylthiophene), both for the linear[119] and cross-linked[118] forms. Neither additives nor a large excess of metallating agent are required, but for the cross-linked material the degree of substitution is critically dependent on solvent and, for copolymers with styrene, on the precise copolymer composition.[120] Mercuration of poly(2-vinylthiophene) is also possible; the reaction proceeds much more easily than the corresponding reaction using polystyrene, high degrees of substitution being obtained using very mild conditions.[121]

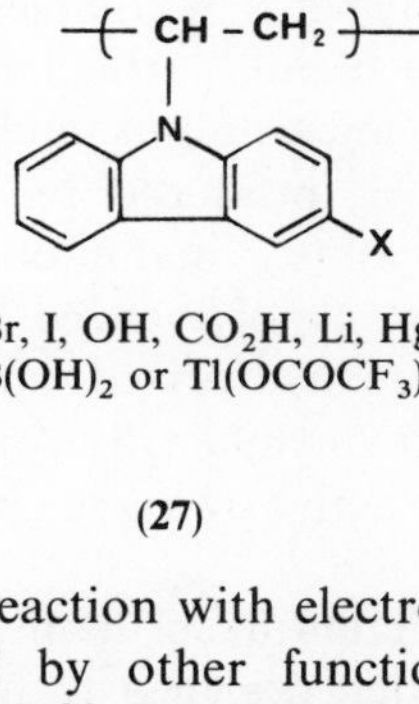

X = Br, I, OH, CO_2H, Li, HgCl
$B(OH)_2$ or $Tl(OCOCF_3)_2$

(**27**)

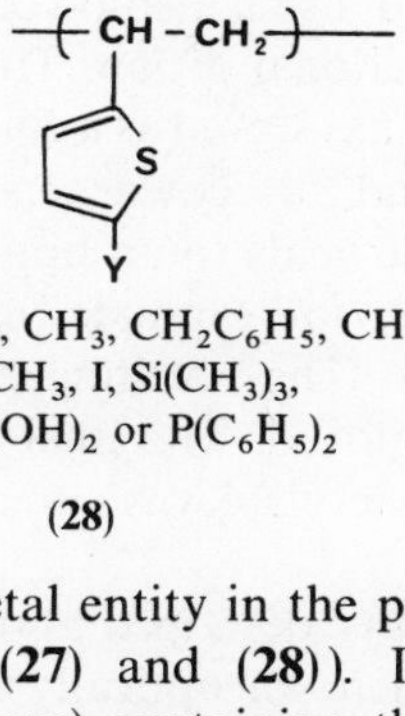

Y = D, CH_3, $CH_2C_6H_5$, CHO,
SCH_3, I, $Si(CH_3)_3$,
$B(OH)_2$ or $P(C_6H_5)_2$

(**28**)

By further reaction with electrophiles, the metal entity in the polymers can be replaced by other functional groups ((**27**) and (**28**)). It has been demonstrated[120] that the poly(2-vinylthiophene) containing the diphenylphosphine residues can be used as an effective reagent for the conversion of alcohols into alkyl halides, i.e. its chemical behaviour parallels that of the analogous reagent supported on polystyrene.[109]

5.3.3 *Miscellaneous systems*

Only the more common polymer systems have been described in the above sections, and an attempt has been made to distinguish between backbone and pendant group modifications. Some reactions are more difficult to catagorize. For example, the aromatic system in polyacenaphthylene (**29**) can be viewed as a main-chain or a pendant entity. This polymer can be brominated[122] or chemically modified via mercuration or thalliation to produce polymers

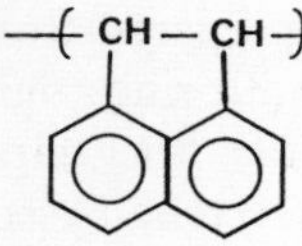

$-(-CH-CH-)-$
(two fused/adjacent phenyl rings)

(29)

containing iodo, phenolic or boronic acid residues,[123] using procedures similar to those described for polystyrene and poly(N-vinylcarbazole). Other polymers are similar to poly(vinyl alcohol) in that they may easily undergo reaction in the main chain or pendant group. For example, polyphosphazenes can be prepared by replacement of the chlorine atoms in polydichlorophosphazenes by nucleophiles such as amines or aryloxides (Scheme 5.29), and further modification of the substituent is possible.[124]

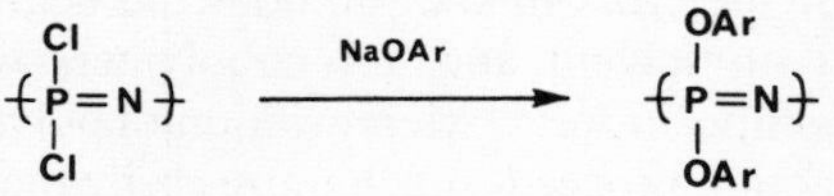

$$\begin{array}{c} Cl \\ | \\ -(-P=N-)- \\ | \\ Cl \end{array} \quad \xrightarrow{NaOAr} \quad \begin{array}{c} OAr \\ | \\ -(-P=N-)- \\ | \\ OAr \end{array}$$

Scheme 5.29

A slightly unusual method of modifying a polymer is by complete removal of a bulky functional group. This is potentially useful, particularly when the bulky group has served as a chiral template. An interesting example is provided by Wulff and his coworkers.[125] A monomer, produced by linking two styrylboronic acids to an optically active polyol (e.g. mannitol) via the diol functions (as, for example, in (20)) was first copolymerized with methyl methacrylate. The resulting polymer was then hydrolysed, and the chiral template removed. The final copolymer of p-vinylphenylboronic acid and methyl methacrylate was found to be optically active owing to chirality in the main chain.

The chirality described above is rather different from the more common situation where an optically active entity, often designed to function as a ligand in the preparation of catalysts for asymmetric syntheses, is linked to a pendant group. Although such materials are frequently made by polymerizing a monomer containing the chiral substituent,[126] they can also be made by functionalizing a preformed polymer. For example, N-methylephedrine has

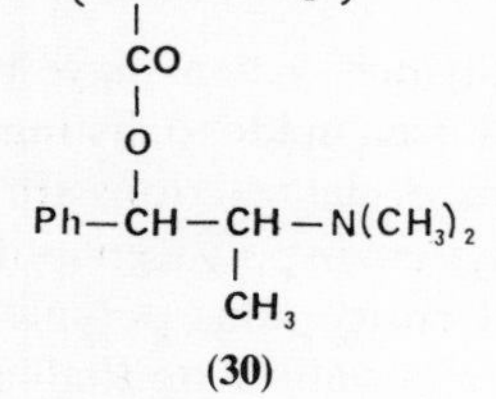

$$\begin{array}{c} -(-CH-CH_2-)- \\ | \\ CO \\ | \\ O \\ | \\ Ph-CH-CH-N(CH_3)_2 \\ | \\ CH_3 \end{array}$$

(30)

been reacted with poly(acryloyl chloride) to give a support (**30**) used in the resolution of mandelic acid.[127]

5.3.4 Concluding remarks

For most types of preformed polymers it is relatively easy to carry out extensive chemical changes. However, the products are usually random copolymers, since it remains difficult to determine precisely all the reaction sites. The examples of chemical modifications given in this short review are relatively recent, and have been chosen to illustrate the very varied chemistry of the more common polymers. However, most of the reactions quoted are merely the first stages in the synthesis of more complex materials or reagents. It is unfortunately beyond the scope of this chapter to describe in detail the applications and further reactions of the materials.

References

1. Hodge, P. and Sherrington, D.C. (Eds), *Polymer-Supported Reactions in Organic Synthesis*, Wiley, Chichester (1980).
2. Sherrington, D.C. and Hodge, P. (Eds), *Synthesis and Separations using Functional Polymers*, Wiley, Chichester (1988).
3. Mathur, N.K., Naray, C.K. and Williams, R.E. *Polymers as Aids in Organic Chemistry*, Academic Press, New York (1980).
4. Akelah, A. and Sherrington, D.C. *Chem. Rev.* **81**, 557 (1981).
5. Ford, W.T. (Ed.), *Polymeric Reagents and Catalysts*, ACS, Washington DC (1986).
6. Kydonieus, A.F. (Ed.), *Controlled Release Technologies, Theory and Applications*, CRC Press, Boca Raton, (1980).
7. Gavina, F., Costero, A.M. and Luis, S.V. *J. Org. Chem.* **49**, 4617 (1984).
8. Marechal, E., Chemical Modification of Synthetic Polymers, in *Comprehensive Polymer Science* (Eds, Allen, G. and Bevington, J.C.), Pergamon Press, Oxford, Vol. 6, p. 1 (1989).
9. Carraher, C.E. and Tsuda, T. (Eds), *Modification of Polymers*, ACS, Washington DC (1980).
10. Carraher, C.E. and Moore, J.A. *Modification of Polymers*, Plenum Press, New York (1983).
11. Martuscelli, E., Marchetta, C. and Nicolais, L. (Eds), *Future Trends in Polymer Science and Technology*, Technomic, Lancaster/Basel (1987).
12. Carraher, C.E., Sheats, J.E. and Pittman, C.U. *Organometallic Polymers 2*, Academic Press, New York (1978).
13. Bevington, J.C. and Thorpe, F.G. *Makromol. Chem., Macromol. Symp.* **20/21**, 1 (1988).
14. Calmon, C. *Reactive Polym.* **1**, 3 (1982).
15. Alami, S.W. and Caze, C. *Eur. Polym. J.* **23**, 883 (1987).
16. Guyot, A., Graillat, C. and Bertholin, M. *J. Mol. Cat.* **3**, 39 (1977/78).
17. Guyot, A., in Reference 2, p. 1.
18. Zussman, M.P. and Tirrell, D.A. *Macromolecules* **14**, 1148 (1981).
19. Zussman, M.P. and Tirrell, D.A. *Polym. Bull.* **7**, 439 (1982).
20. Baines, F.C. and Bevington, J.C. *J. Polym. Sci., Part A-1* **6**, 2433 (1968).
21. Ford, W.T., Mohanraj, S. and Periyasamy, M. *Brit. Polym. J.* **16**, 179 (1984).
22. Andreis, M. and Koenig, J.L. *Adv. Polym. Sci.* **89**, 69 (1989).
23. Defieuw, G., Groeninckx, G. and Reynaers, H. *Polymer* **30**, 595 (1989).
24. Trumbo, D.L. and Marvel, C.S. *J. Polym. Sci., Polym. Symp.* **74**, 45 (1986).
25. Ang, C.H., Garnett, J.L., Levot, R.G. and Long, M.A. in Reference 10, p. 33.
26. Potapov, V.K., Turkin, S.I., Veiko, V.P., Shabarova, Z.A. and Prokf'ev, M.A. *Doklady Chem.* **241**, 405 (1978).

27. Kuntz, I. and Hudson, B.E. in Reference 10, p. 53.
28. See Reference 8, p. 7.
29. Lagadet, F., Deffieux, A., Servens, C. and Fontanille, M. *Eur. Polym. J.* **25**, 63 (1989).
30. Pinazzi, C., Vassant, J. and Reyx, D. *Eur. Polym. J.* **13**, 395 (1977).
31. Tessier, M. and Marechal, E. *Eur. Polym. J.* **20**, 269 (1984).
32. Siddiqui, S. and Cais, R.E. *Macromolecules* **19**, 59 (1986).
33. Sanui, K., MacKnight, W.J. and Lenz, R.W. *Macromolecules* **7**, 952 (1974).
34. Kershaw, J.R., Edmond, L. and Rizzardo, E. *Polymer* **30**, 361 (1989).
35. Naqui, M.K. *J. Macromol. Sci., Rev. Macromol. Chem. Phys.* **C25**, 119 (1985).
36. Vorenkkamp, F.J. and Challa, G. *Polymer* **29**, 86 (1988).
37. Okawara, M. and Ochiai, Y. in Reference 9, p. 41.
38. Nkansah, A. and Levin, G. in Reference 10, p. 109.
39. Pelter, M.W. and Taylor, R.T. *J. Polym. Sci., Polym. Chem. Ed.* **26**, 2651 (1988).
40. Huth, J.A. and Danielson, N.D. *Anal. Chem.* **54**, 930 (1982).
41. Dandos, A. and Rempp, P. *Compt. Rend.* **254**, 1064 (1962).
42. Dias, A.J. and McCarthy, T.J. *Macromolecules* **18**, 1826 (1985).
43. Rose, J.B. *Polymer* **15**, 456 (1974).
44. Johnson, B.C., Yilgor, I., Lloyd, D.R. and McGrath, J.E. *J. Polym. Sci., Polym. Chem. Ed.* **22**, 721 (1984).
45. Noshay, A. and Robeson, L.M. *J. Appl. Polym. Sci.* **20**, 1885 (1976).
46. Percec, V. and Auman, B.C. *Polym. Prepr.* **25**, 122 (1984).
47. Willson, C.G., Frechet, J.M. and Farrell, M.J. in Reference 10, p. 25.
48. Marayana, K., Takeuchi, K. and Tanizaki, Y. *Polymer* **30**, 476 (1989).
49. Chanda, M., O'Driscoll, K.F. and Rempel, G.M. *Reactive Polym.* **5**, 157 and 225 (1987).
50. Lee, C.S. and Daly, W.H. *Adv. Polym. Sci.* **15**, 61 (1979).
51. Bochkin, A.M. and Pomogailo, A.D. *Reactive Polym.* **9**, 99 (1988).
52. Norisawa, M., Ono, K. and Murakami, K. *Polymer* **30**, 1540 (1989).
53. Sinton, S.W. *Macromolecules* **20**, 2430 (1987).
54. Shibayama, M., Sato, M., Kimura, Y., Fujiwara, H. and Nomura, S. *Polymer* **29**, 336 (1988).
55. Bowcher, T.L. and Dawber, J.G. *Polymer* **30**, 215 (1989).
56. Harris, F.W. in Reference 6, p. 63.
57. Gebelein, C.G. and Burnfield, K.E. in Reference 9, p. 83.
58. Matsunaga, T. and Ikada, Y. in Reference 9, p. 390.
59. Johnson, A., Klesper, E. and Wirthlin, T. *Makromol. Chem.* **177**, 2397 (1976).
60. Barth, V. and Klesper, E. *Polymer* **17**, 777, 787 and 893 (1976).
61. See Reference 13, p. 5.
62. Kheradmand, H., Francois, T. and Plazanat, V. *Polymer* **29**, 860 (1988).
63. Gold, V., Liddiard, C.J. and Martin, J.L. *J. Chem. Soc., Faraday Trans. I* **73**, 1119 (1977).
64. Rempp, P. *Pure and Appl. Chem.* **46**, 9 (1976).
65. Bourgignon, J.T. and Galin, J.C. *Macromolecules* **10**, 804 (1977).
66. Galin, J.C. in Reference 9, p. 119.
67. Chung, T.C., Schlesinger, Y., Etemad, H.S., Macdiarmid, A.G. and Heeger, A.J. *J. Polym. Sci., Polym. Phys. Ed.* **22**, 1239 (1984).
68. Babu, G.N., Deshpande, A., Dhal, P.K. and Deshpande, D.D. in Reference 10, p. 65.
69. Kahovec, J. and Coupek, J. *Reactive Polym.* **8**, 105 (1985).
70. Yeager, W.L. and Castelli, V.J. in Reference 12, p. 175.
71. Akelah, A. *Reactive Polym.* **8**, 273 (1988).
72. George, B.K. and Pillai, V.N. *Polymer* **30**, 178 (1989).
73. Domb, A.J., Cravalho, E.G. and Langer, R. *J. Polym. Sci., Polym. Chem. Ed.* **26**, 2623 (1988).
74. Shigetomi, Y., Hori Y. and Kojima, T. *Bull. Chem. Soc. Jap.* **53**, 1475 (1980).
75. Poinescu, I., Vlad, C.D. and Carpov, A. *Reactive Polym.* **2**, 261 (1984).
76. Merrifield, R.B. *J. Amer. Chem. Soc.* **82**, 2149 (1963).
77. Farrell, M.J. and Frechet, J.M.J. *J. Org. Chem.* **41**, 3877 (1976).
78. Frechet, J.M.J., Nuyens, L.J. and Seymour, E. *J. Amer. Chem. Soc.* **101**, 432 (1979).
79. Relles, H.M. and Schluenz, R.W. *J. Amer. Chem. Soc.* **96**, 6469 (1974).
80. Hodge, P. in Reference 1, p. 477.
81. Sherrington, D.C. in Reference 1, p. 1.
82. Kusama, T. and Hayatsu, H. *Chem. Pharm. Bull.* **18**, 319 (1970).

83. Warshawky, A., Deshe, A., Rossey, G. and Patchornik, A. *Reactive Polym.* **2**, 301 (1984).
84. Haridasan, V.K., Ajayaghosh, A. and Pillai, V.N.R. *J. Org. Chem.* **52**, 2662 (1987).
85. Menger, F.M. and Tsuno, T. *J. Amer. Chem. Soc.* **111**, 4903 (1989).
86. Elman, B. and Moberg, C. *J. Organomet. Chem.* **294**, 117 (1985).
87. Alexandratos, S.D. and Bates, M.E. *Macromolecules* **21**, 2905 (1988).
88. Tomoi, M., Ishigaki, S., Artita, Y. and Kakiuchi, H. *J. Polym. Sci., Polym. Chem. Ed.* **26**, 2251 (1988).
89. Sahni, S.K. and Reedijk, J. *Co-ord. Chem. Rev.* **59**, 1 (1984).
90. Sreekumar, K. and Pillai, V.N.R. *Polymer* **28**, 1599 (1987).
91. Antonsson, T. and Moberg, C. *Reactive Polym.* **8**, 113 (1988).
92. Fyles, T.M. and Leznoff, C.C. *Can. J. Chem.* **54**, 935 (1976).
93. Grubbs, R.H. and Su, S-C.H. *J. Organomet. Chem.* **122**, 151 (1976).
94. Chan, T.H. and Huang, W.Q. *J. Chem. Soc., Chem. Commun.* 909 (1985).
95. Zhangyu, Z., Hungwen, H. and Tsi-Yu, K. *Reactive Polym.* **9**, 249 (1988).
96. Evans, D.C., George, M.H. and Barrie, J.A. *J. Polym. Sci., Polym. Chem. Ed.* **12**, 247 (1974).
97. Plate, N.A., Jampolskaya, M.A., Davydova, S.L. and Kargin, V.A. *J. Polym. Sci., Part C* **22**, 547 (1969).
98. Chalk, A.J. *J. Polym. Sci., Polym. Lett. Ed.* **6**, 649 (1968).
99. Evans, D.C., Phillips, L., Barrie, J.A. and George, M.H. *J. Polym. Sci., Polym. Lett. Ed.* **12**, 19 (1974).
100. Chalk, A.J. and Hoogeboom, T.J. *J. Organomet. Chem.* **11**, 615 (1968).
101. Frechet, J.M.J., Rolls, W. and Darling, G. in Reference 11, p. 128.
102. Appel, R. and Willms, L. *Chem. Ber.* **114**, 858 (1981).
103. McKinley, S.V. and Rakshys, J.W. *J. Chem. Soc., Chem. Commun.* 134 (1972).
104. Grubbs, R.H. and Su, S-C.H. in Reference 12, p. 129.
105. Garrou, P.E. and Bates, B.C. in Reference 12, p. 123.
106. Schwartz, R.H. and San Fillipo, J. *J. Org. Chem.* **44**, 2705 (1979).
107. Hodge, P., Khoshdel, E. and Waterhouse, J. *Makromol. Chem.* **185**, 489 (1984).
108. Bernard, M. and Ford, W.T. *J. Org. Chem.* **48**, 326 (1983).
109. Hodge, P. and Richardson, G. *J. Chem. Soc., Chem. Commun.* 622 (1975).
110. Bullen, N.P., Hodge, P. and Thorpe, F.G. *J. Chem. Soc., Perkin Trans. I*, 1863 (1981).
111. Lenfeld, J., Peska, J. and Stamberg, J. *Reactive Polym.* **1**, 47 (1982).
112. Janout, V. and Regen, S.L. *J. Org. Chem.* **47**, 2213 (1982).
113. Hrudkova, H., Cefelin, P. and Janout, V. *Polym. Bull.* **20**, 143 (1988).
114. Stevenson, T.A. and Taylor, R.T. *Reactive Polym.* **8**, 7 (1988).
115. Wielema, T.A. and Engberts, J.B.F.N. *Eur. Polym. J.* **23**, 947 (1987).
116. Lyons, A.M., Vasile, M.J., Pearce, E.M. and Waszczak, J.V. *Macromolecules* **21**, 3125 (1988).
117. Frechet, J.M.J. and Vivas de Meftahi, M. *Brit. Polym. J.* **16**, 193 (1984).
118. Al-Kadhumi, A.A.H.A., Hodge, P. and Thorpe, F.G. *Brit. Polym. J.* **19**, 325 (1987).
119. Al-Kadhumi, A.A.H.A., Hodge, P. and Thorpe, F.G. *Polymer* **26**, 1695 (1985).
120. Hodge, P., Liu, M-G. and Thorpe, F.G. *Polymer* **31**, 140 (1990).
121. Al-Kadhumi, A.A.H.A., Hodge, P. and Thorpe, F.G. *Brit. Polym. J.* **16**, 225 (1984).
122. Hodge, P., Hunt, B.J. and Shakshier, I.H. *Polymer* **26**, 1701 (1985).
123. Al-Kadhumi, A.A.H.A., Hodge, P., Hunt, B.J. and Thorpe, F.G. *Reactive Polym.* **7**, 15 (1987).
124. Allcock, H.R., Mang, M.N., Dembek, A.A. and Wynne, K.J. *Macromolecules* **22**, 4179 (1989).
125. Wulff, G. and Hohn, J. *Macromolecules* **15**, 1255 (1982).
126. Stille, J.K. *J. Macromol. Sci. Chem.* **A21**, 1689 (1984).
127. Blaschke, G. *Chem. Ber.* **107**, 237 (1974).

6 Terminally reactive oligomers: telechelic oligomers and macromers

J.R. EBDON

6.1 Introduction

Oligomers are short-chain polymers; the name originates from the Greek words 'oligo', meaning few, and 'mer', meaning part. The dividing line between oligomers and polymers is a somewhat arbitrary one but is best considered as corresponding to a region beyond which the bulk physical and mechanical properties of the polymer no longer depend significantly on chain-length. This region will vary from polymer to polymer depending upon the chemical complexity of the repeat unit but corresponds typically to a degree of polymerization in the range 10–100.

Considerable and mounting interest has been shown, particularly over the last decade, in the preparation of oligomers with terminal functional groups capable of further reaction to give new types of polymer not accessible by conventional polymerizations, e.g. block and graft copolymers. Reactive oligomers also lend themselves to processing methods and applications for which conventional high molecular weight polymers are less suitable, e.g. in reactive injection moulding (RIM) in which chain extension and/or cross-linking takes place during a moulding operation, and in the formulation of liquid surface coating systems with high solids contents for which high molecular weight polymers would give unacceptably high solution viscosities.

Terminally reactive oligomers can be divided into two main categories: telechelic oligomers and macromers (often more informatively referred to as macromonomers). The adjective 'telechelic' was proposed by Uranek et al. in 1960 to describe oligomers having two reactive end-groups.[1] The term is derived from the Greek words 'tele, meaning at a distance and 'chelos', meaning claw. The end-groups of a telechelic oligomer can take part in reactions with the end-groups of another telechelic oligomer or with a difunctional small molecule to give block polymers (equations 6.1 and 6.2). Typical end-groups include hydroxyl, thiol, halide, carboxyl and amine.

The name 'macromer' (meaning macromolecular monomer) was coined by Milkovich in 1974[2] specifically to describe some oligomers of styrene prepared

$$n \text{ A}\sim\sim\sim\text{B} + n \text{ A}'\sim\sim\sim\text{B}' \longrightarrow +\!\!\{\text{A}\sim\sim\sim\text{BA}'\sim\sim\sim\text{B}'\}_n$$

(6.1)

$$nA\sim\sim\sim B + nA'-B' \longrightarrow (A\sim\sim\sim BA'-B')_n \quad (6.2)$$

by anionic polymerization and having terminal vinyl groups capable of addition polymerization, but is now applied more widely to cover any oligomer having at least one homo- and/or copolymerizable end-group, e.g. a terminal vinyl, acrylic or acetylenic group. Oligomers can now be made by a wide variety of polymerization processes (addition and stepwise), and the reactive end-group(s) can be introduced during the oligomerization or in a second stage and can be designed to be polymerizable by either an addition (equation 6.3) or a stepwise mechanism (equation 6.4).

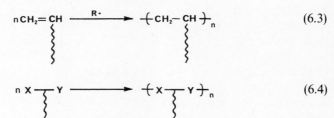

Thus macromers are now available with a variety of chemical structures lending themselves to the preparation of polymeric materials with a variety of applications. Macromers with one polymerizable end-group will produce graft or comb polymers when the end-groups are homopolymerized or, as is more commonly done, copolymerized with a conventional low molecular weight monomer, whereas macromers with two polymerizable end-groups will form networks when the end-groups are polymerized.

As more and more terminally reactive oligomers are synthesized and more and more reactions of them are demonstrated, it becomes apparent that the distinction between a macromer and a telechelic oligomer is not necessarily obvious from its structure because it is possible for a terminally reactive oligomer to be both a macromonomer and a telechelic oligomer. For example, consider the possible reactions between an oligomer with a terminal maleimido unit and one with a terminal dienyl unit. Simply heating a mixture of the oligomers might result in a Diels–Alder reaction between the end-groups to form a block copolymer, whereas heating in the presence of a radical initiator could lead to a copolymerization of the end-groups to give a comb polymer (Scheme 6.1). Similar ambiguity could arise over oligomers possessing terminal vinyl ketone or acrylic groups which could behave as telechelic oligomers when the vinyl groups take part in conjugate addition (Michael and Mannich) reactions and as macromers when the groups are simply polymerized (Scheme 6.2). Thus it may be more satisfactory now to define macromers and telechelic oligomers by reference to whether or not the end-groups are behaving as monofunctional end-groups (telechelic oligomers) or as difunctional, and therefore polymerizable, end-groups (macromers) in further reactions with other oligomers or with low molecular weight reactants.

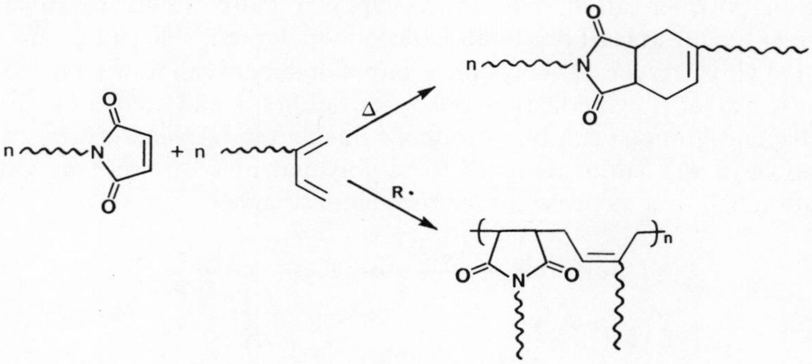

Scheme 6.1

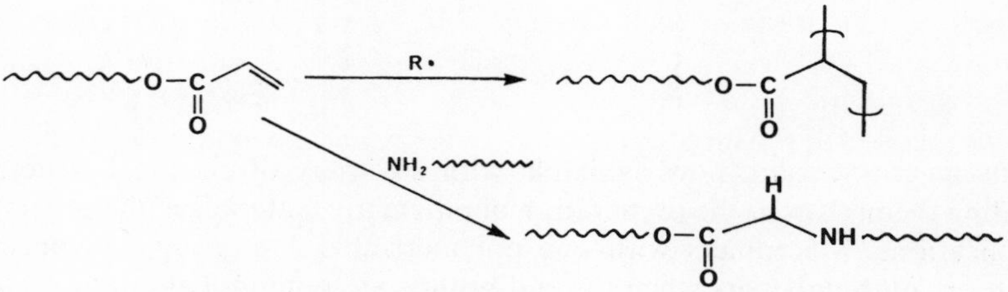

Scheme 6.2

The sections which follow review the methods now available for synthesizing telechelic oligomers and macromers with various types of end-group, consider the reactivities of the materials so formed and outline some of their more important applications.

6.2 The synthesis of terminally reactive oligomers

It is important that terminally reactive oligomers should have the desired numbers (one or two) and types of end-groups and that each oligomer molecule should be identical to its brethren in these respects. For some applications, e.g. in the preparation of block and graft polymers with well-defined structures for model studies, it may also be important for the oligomer sample to have a narrow molecular weight distribution and for the chains to have a particular and regular stereochemistry. These requirements indicate that, where practicable, it is advantageous to prepare the oligomer by a polymerization mechanism offering the greatest control over the initiation, propagation and termination steps. Such control is offered best by 'living' anionic, cationic and group transfer (GTP) mechanisms in which the desired end-group(s) can be introduced during either the initiation or the termination steps. However, the preparation of only a relatively few types of terminally reactive oligomer can be accomplished using these mechanisms and therefore other mechanisms of polymerization have been exploited also, including radical and coordination addition polymerization and many types of stepwise (condensation and rearrangement) polymerization. The preparation of terminally functionalized oligomers has been regularly and extensively reviewed, including quite recently, and therefore only general principles and a selection of illustrative examples are given here.[3-7]

6.2.1 By anionic polymerization

6.2.1.1 *Telechelic oligomers* The usual strategy for synthesizing telechelic oligomers by anionic polymerization is to initiate polymerization with a suitable anionic initiator used at a relatively high concentration (relative to monomer) to keep chain lengths short, and then to terminate the resulting living oligomeric anion with a terminating agent which introduces the required end-group. Since telechelic oligomers are often required to have two monofunctional end-groups of the same type, it is particularly advantageous to initiate the polymerization with a difunctional anionic initiator such as sodium naphthalene. However, the majority of reactions for introducing functional end-groups into anionically prepared polymers during termination have so far been demonstrated with polymers initiated with monofunctional initiators such as butyl lithium.

Hydroxy end-groups can be introduced into living polystyrenes and polydienes by reacting the living polyanion with ethylene oxide to give an alkoxide ion followed by immediate quenching with water or an alcohol.[8-10] Alternatively, formaldehyde can be used as a terminating agent.[11]

Primary amino end-groups have been introduced into living polymers with blocked functional terminating agents which are subsequently hydrolysed to restore the amine functionality.[12] Secondary amino end-groups can be introduced via titration with N-alkyl aziridines whilst reaction with α,ω-alkylene chloramines gives tertiary amino end-groups (Scheme 6.3).[13]

Carboxy end-groups are most easily produced by reaction of the living polyanion with carbon dioxide[14] although significant quantities of dimeric ketones and trimeric carbinols can also result. However, almost quantitative conversion of living ends to carboxylate groups can be achieved if the living polymers are first converted to less reactive Grignard reagents.[15] Carboxylate end-groups can also be introduced by reactions of living polymers with lactones and cyclic anhydrides equation (6.5).[16]

Halide end-groups can be generated by reaction of the living polymer with excess halogen although the reaction is not quantitative owing to Wurtz

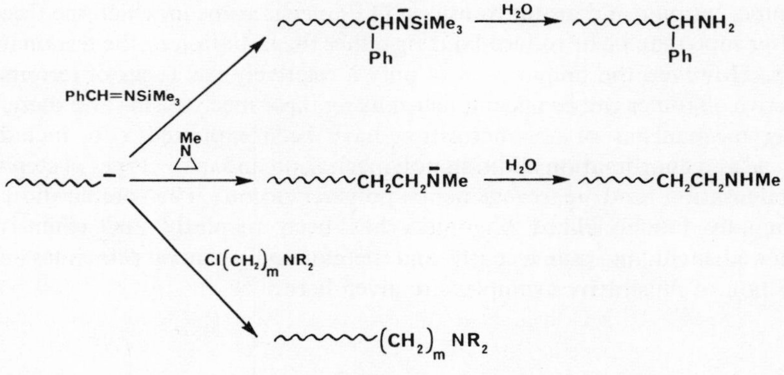
H₂O
CHNSiMe₃
CHNH₂
Ph
Ph
PhCH=NSiMe₃
Me
N
H₂O
CH₂CH₂NMe
CH₂CH₂NHMe
Cl(CH₂)m NR₂
(CH₂)m NR₂

Scheme 6.3

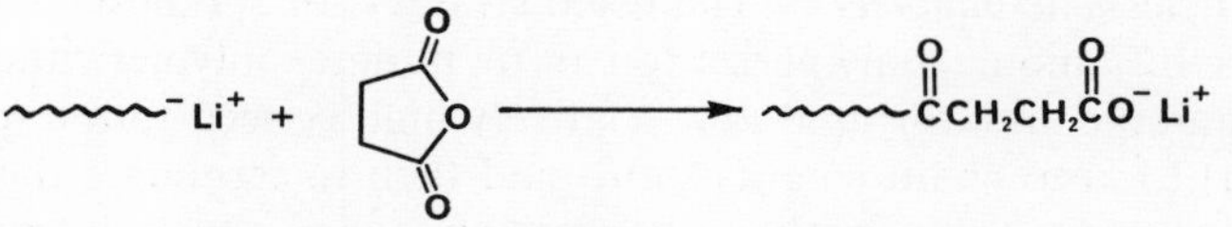
Li⁺ +
O
O
O
CCH₂CH₂CO⁻ Li⁺
(6.5)

coupling.[17,18] Once again there are advantages in first converting the living polymer to a less reactive Grignard reagent. An alternative method of introducing halide end-groups is to react the living polymer with an excess of an α,ω-dihalide, e.g. 1,4-bis(bromomethyl)benzene (equation 6.6).[19,20]

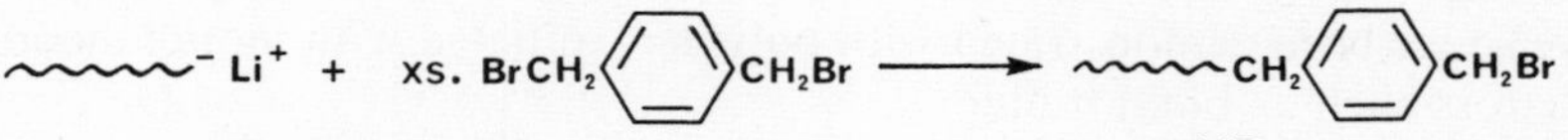
Li⁺ + xs. BrCH₂
CH₂Br
CH₂
CH₂Br
+LiBr
(6.6)

Reactive end-groups can also be introduced into living anionic polymers at the initiation stage by using a suitably functionalized initiator. The simplest example is the use of potassium amide in liquid ammonia to produce polystyrenes having amine end-groups.[21] Usually, however, the functional group on the initiator has to be of a protected type since many of the end-groups required in telechelics are reactive towards anions. For example, the organolithium initiators (**1**) and (**2**) have been used to introduce hydroxy and amino functionality respectively.[22,23]

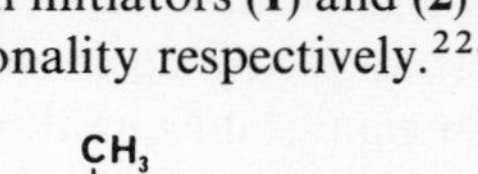
CH₃
EtOCHO(CH₂)₅CH₂⁻ Li⁺

(1)

(Me₃Si)₂N
Li⁺

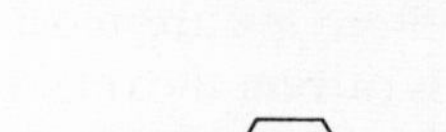

(2)

The examples considered above are of telechelics prepared from the anionic polymerization of vinyl and divinyl (diene) monomers. However, many cyclic monomers also can be polymerized anionically including oxiranes (e.g.

ethylene oxide), cyclic sulphides, lactones and cyclic siloxanes. Anionic polymerization of oxirane initiated by sodium hydroxide gives, after reaction with a proton donor, oligoethylene oxides with terminal hydroxyl groups.[24] Similar reactions occur also with substitued oxiranes and can be used, for example, to prepare telechelic poly(propylene oxide)s. An alternative initiator for these polymerizations is sodium naphthalene, use of which will generate double-ended growing chains.[25,26] Propylene sulphide has been similarly polymerized using sodium naphthalene as initiator with termination by 1-chloromethyl naphthalene to produce dinaphthalene-ended telechelics.[27]

The anionic polymerization of β-lactones can be readily initiated using tertiary amines. If α,α-disubstituted lactones are used (e.g. pivalolactone (3)), polymerization proceeds without transfer to give telechelics with amine and carboxylate end-groups.[28-30] Living polymers from ε-caprolactone convertible to telechelics can be prepared using complex alkoxide initiators such as $(BuO)_4Al_2O_2Zn$.[31,32]

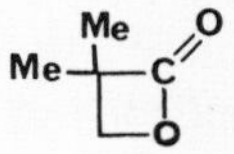

(3)

Cyclic siloxanes can be polymerized either cationically or anionically to give linear polysiloxanes. Anionic initiators include hydroxides, alkoxides, phenolates and silanolates. Bifunctional polysiloxanes with a variety of end-groups can be obtained using appropriately functionalized 'end-blockers'.[33-35]

6.2.1.2 *Macromers* Macromers can be obtained from living anionic polymerizations by first converting the living polymer to a telechelic using one of the above methods and then further reacting the terminal functional group(s) to introduce polymerizable unsaturation. However, it is rather more elegant to introduce the unsaturation using a suitably functionalized initiator or terminating agent.

Living polystyrenes, for example, have been converted to macromers capable of addition polymerization by termination, either directly or after reacting the living anion with oxirane to give a less reactive alkoxide ion, with a variety of unsaturated alkyl or acyl halides, e.g. (4), (5) and (6).[36-41] The prior conversion of the living carbanion to an alkoxide is particularly recommended with the use of methacryloyl chloride (5) to prevent attack of the anion on the conjugated unsaturation.[37,38] Strategies similar to the above have been used

$CH_2=CHCH_2Br$ $CH_2=C(Me)COCl$ $CH_2=CH\text{-}C_6H_4\text{-}CH_2Cl$

(4)   (5)   (6)

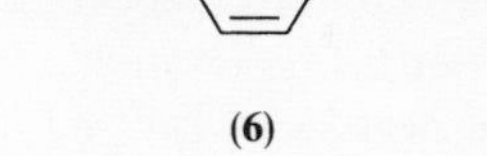

also to prepare macromers from vinyl pyridines,[42-44] ethylene oxide,[44-47] methyl methacrylate[48-51] and dimethylsiloxanes[52]

Introduction of unsaturation at the initiation stage in anionic polymerizations of styrene and ethylene oxide has been accomplished using a variety of suitably functionalized lithium and potassium salts, e.g. (7), (8) and (9).[45,53,54] Methyl methacrylate macromers also have been prepared by this technique using the functionalized Grignard reagent, o-vinylbenzylmagnesium chloride.[55] Recently, some dimethylsiloxane macromers have been produced with the aid of partly lithiated alkenylsiloxanol initiators.[56,57]

$$CH_2=CH^-Li^+ \qquad CH_2=CHCH_2^-Li^+ \qquad CH_2=C(Me)\text{-}C_6H_4\text{-}CH_2O^-K^+$$

(7)　　　　　　(8)　　　　　　　　(9)

6.2.2 By group transfer polymerization

6.2.2.1 Telechelic oligomers
Whilst living anionic polymerization provides a satisfactory route to many vinyl and dienyl oligomers it is less satisfactory for α,β-unsaturated carbonyl compounds, e.g. acrylic and methacrylic monomers, because of the potential for side reactions arising from attack of the anion on the carbonyl group and, in the case of methacrylic monomers, from abstraction of an α-proton.

Comparatively recently, however, it has been demonstrated that acrylic and methacrylic monomers, provided they contain no active hydrogens, can be polymerized to living polymers without significant side reactions by what has come to be known as group transfer polymerization (GTP). This type of polymerization, which is described in detail in Chapter 2, is typically initiated by silyl ketene acetals in the presence of catalytic amounts of weakly nucleophilic anions or of Lewis acids.

Terminally reactive oligomers can be prepared by GTP by strategies akin to those used in anionic polymerizations, i.e. using suitably functionalized initiators and/or terminating agents. For example, α,ω-hydroxy polyacrylates and methacrylates have been obtained by initiating polymerization with an initiator containing a protected hydroxy group and then coupling the resulting living polymers with 1,4-bis(bromomethyl)benzene. The hydroxy groups are deprotected in a second stage to introduce the terminal functionality (Scheme 6.4).[58-60] The difunctional initiator, (10), has been used in a similar fashion to produce polymer with α,ω-dicarboxy functionality.[59-61] The living ends of group transfer polymers can be

$$\begin{matrix} Me \\ Me \end{matrix}\!\!>=<\!\!\begin{matrix} OSiMe_3 \\ OSiMe_3 \end{matrix}$$

(10)

efficiently acylated with a variety of acylating agents; difunctional reagents have recently been used to couple monofunctional living PMMA to produce telechelic diols.[62]

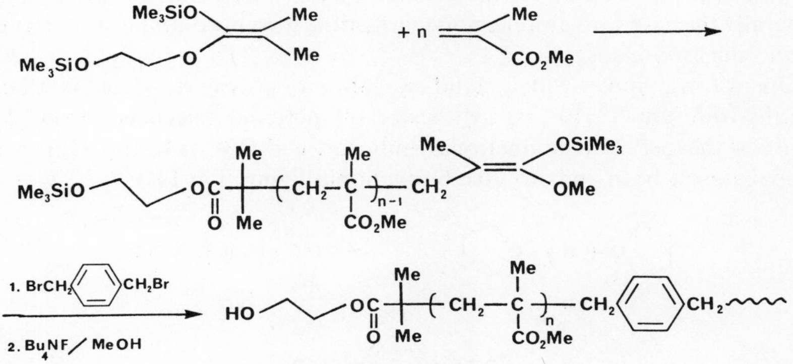

Scheme 6.4

6.2.2.2 *Macromers* As with anionic polymerization, it is possible to convert telechelic oligomers to macromers by further reactions of the terminal functional groups. However, here also it is possible to introduce the terminal unsaturation by more direct methods during initiation and/or termination.

For example, it has been demonstrated that both termination of living poly(methyl methacrylate) with *p*-vinylbenzyl bromide and initiation with the functionalized initiators, vinylphenylketene methyl trimethyl silyl acetal (**11**) and α-trimethyl silyl vinyl benzene cyanide (**12**), give styryl-ended macromers in high yield.

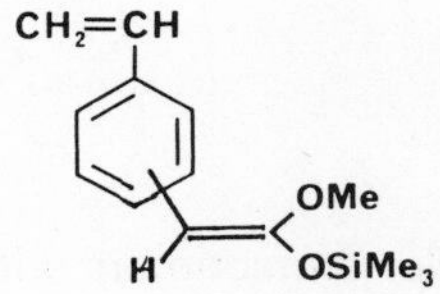

(**11**) 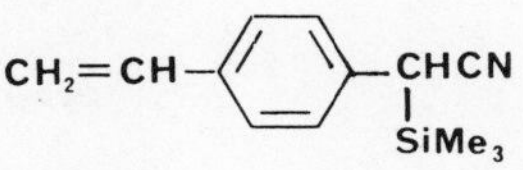   (**12**)

6.2.3 *By cationic polymerization*

6.2.3.1 *Telechelic oligomers* The generally received wisdom is that cationic polymerization does not offer a satisfactory route to many high and low molecular weight polymers with well-defined structures and molecular weights. This is because of the ease with which the growing cationic chain-end can often undergo inter- or intra-molecular transfer or rearrangement. However, over the past decade it has been demonstrated that for several monomers sufficient control of initiation, propagation and termination can be achieved to allow useful telechelic products to be obtained. Although in principle it should be possible to produce telechelic oligomers by cationic polymerization using either monofunctional or bifunctional initiators, as is generally the case with anionic polymerization, it seems that the latter strategy often offers more success.

One of the most widely studied cationic polymerizations is that of tetrahydrofuran (THF). To synthesize bifunctional telechelics from THF requires the use of a bifunctional initiator, e.g. **(13)**, **(14)** and **(15)**, which propagates at both ends to give bifunctional living polyTHF.[65,66]

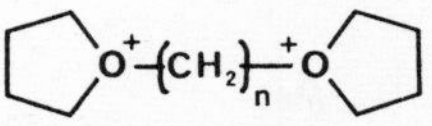

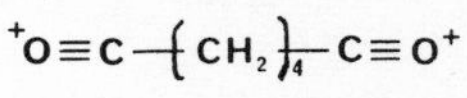

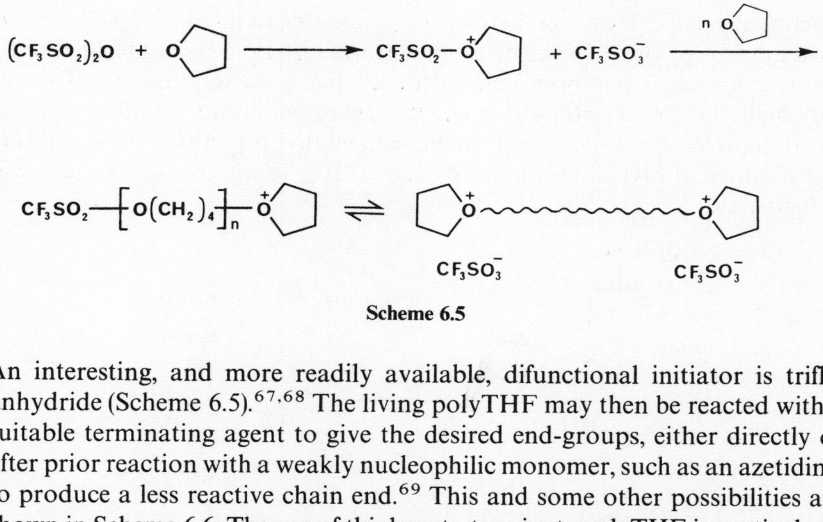

Scheme 6.5

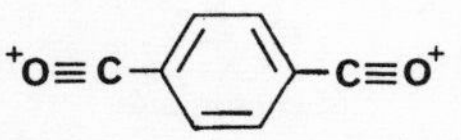

An interesting, and more readily available, difunctional initiator is triflic anhydride (Scheme 6.5).[67,68] The living polyTHF may then be reacted with a suitable terminating agent to give the desired end-groups, either directly or after prior reaction with a weakly nucleophilic monomer, such as an azetidine, to produce a less reactive chain end.[69] This and some other possibilities are shown in Scheme 6.6. The use of thiolane to terminate polyTHF is particularly interesting because it produces telechelics, the end-groups of which are unaffected by water but which may be ring-opened with charged nucleophiles to produce a variety of additional products.[70] Alternatively, the desired end-groups may be introduced during the polymerization of THF by the use of appropriate transfer agents. For example, the polymerization of THF in the presence of organic acid anhydrides leads to polyTHF telechelics with terminal ester groups[71].

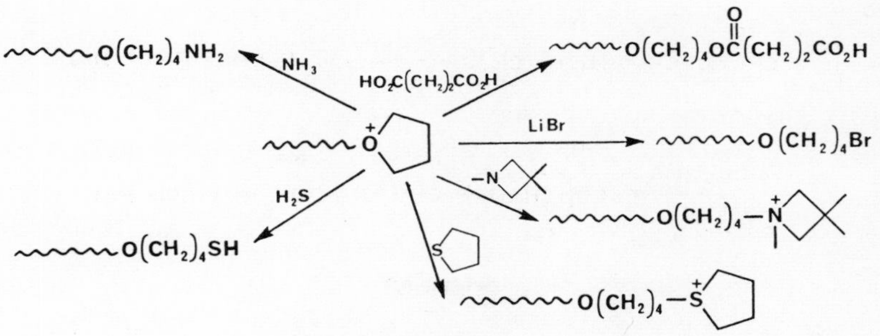

Scheme 6.6

A number of other cyclic monomers also have been polymerized by cationic means to give telechelic products after suitable termination: for example, three-membered cyclic ethers (oxiranes), three- and four-membered cyclic amines (aziridines and azetidines) and some cyclic acetals and siloxanes.[72] However, it should be noted that with the oxiranes, cationic polymerization must be carried out in the presence of an added alcohol by the so-called 'activated monomer' mechanism to prevent intramolecular transfer of the growing end (see Section 1.3.3).[73,74] Telechelics have also been produced recently from 2-alkyl-2-oxazolines by using a bis(2-oxazolinium salt) as a difunctional initiator. Termination of both living ends with water gives an oligoglycol whilst termination with ammonia or an alkylamine gives an oligodiamine.[75] A recent example of the preparation of telechelic siloxanes is that of α,ω-carboxypropyl oligodimethylsiloxanes by the cationic ring-opening polymerization of octamethylcyclotetrasiloxane in the presence of 1,3-bis(carboxypropyl)tetramethyldisiloxane as end-capping agent and triflic acid as initiator.[76]

Carbocationic polymerizations of vinyl monomers do not generally proceed smoothly to give living polymers which can be end-capped to give telechelics with high degrees of end-functionality. However, an interesting exception is the polymerization of vinyl ethers initiated by HI/I_2 (see Section 1.3.4). This particular initiation mechanism gives rise to polymers with very narrow molecular weight distributions which can be quantitatively end-capped to give useful telechelics.[77,78] One variant of the method uses a divinyl ether as a bifunctional initiator; this is then used to polymerize a monovinyl ether to give a double-ended telechelic after suitable end-capping (Scheme 6.7). Alternatively, the required end-functionality can be introduced by using both a functional initiator and a functional terminating agent.[79]

A second useful example of carbocationic polymerization for the preparation of telechelics is the living polymerization of certain olefins, especially isobutene, in the presence of 'inifers' (compounds that act as both initiators and transfer agents). The mechanism has been most extensively explored by Kennedy and coworkers and in its simplest form is represented by

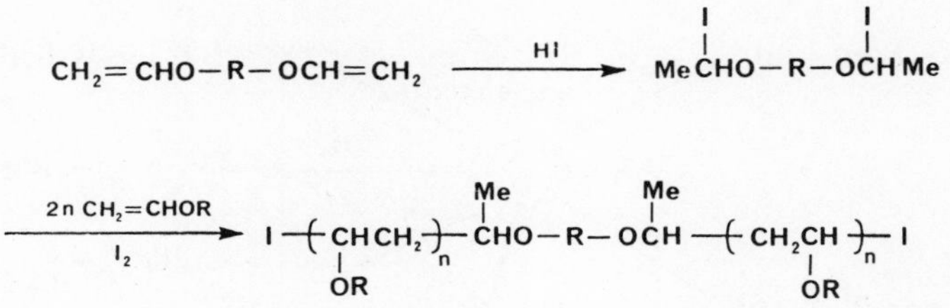

172   NEW METHODS OF POLYMER SYNTHESIS

$$CH_2=CHO-R-OCH=CH_2 \xrightarrow{HI} Me\overset{|}{\underset{|}{C}}HO-R-O\overset{|}{\underset{|}{C}}HMe$$

$$\xrightarrow[I_2]{2n\ CH_2=CHOR} I\underset{\underset{OR}{|}}{-(CHCH_2)_n}\overset{Me}{\underset{|}{-CHO}}-R-O\overset{Me}{\underset{|}{CH}}\underset{\underset{OR}{|}}{-(CH_2CH)_n-I}$$

Scheme 6.7

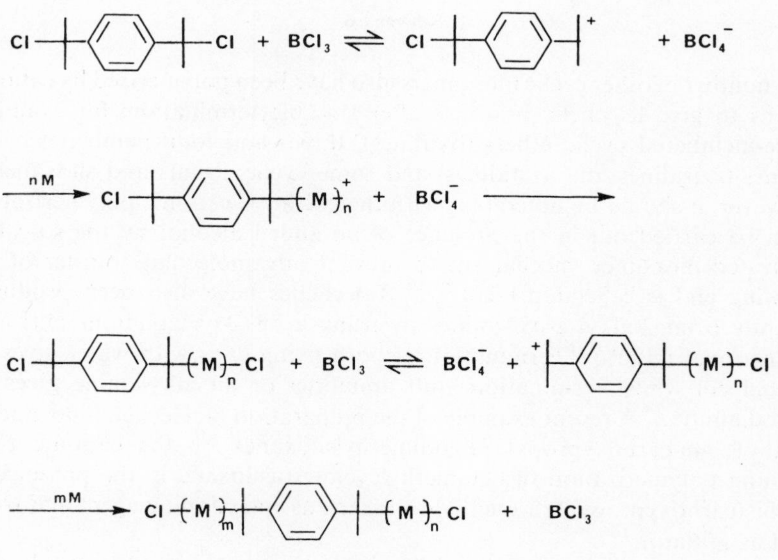

Scheme 6.8

Scheme 6.8.[80] The functionality of the telechelic depends upon the functionality of the inifer; for doubly-ended telechelics, bifunctional inifers are required (binifers). The inifer need not necessarily be an aromatic compound nor need it be a chloride. In more recent work, inifers based on non-aromatic hindered ethers, acetates and alcohols have been demonstrated.[81–84] Lewis acids other than BCl_3 are also effective in promoting polymerization and other olefins, such as β-pinene, can also be converted to telechelics[85].

6.2.3.2 *Macromers* By analogy with anionic methods, terminal unsaturation can be introduced into oligomers prepared by cationic polymerization through the use either of functionalized initiators or of functionalized terminating agents. For example, polyTHF macromers with terminal ethylenic, styrenic and vinyl ketone-type unsaturation have been produced by initiation with **(16)**, **(17)** and **(18)** respectively. [86–88] Styrenic unsaturation has also been introduced into polyTHF by termination with the alkoxides, (19) and (20).[89,90]

$CH_2=CHCH_2^+ \ BF_4^-$ (16)

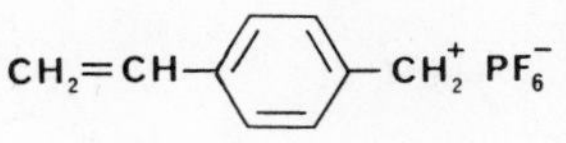

$CH_2=CH-C_6H_4-CH_2^+ \ PF_6^-$ (17)

$CH_2=\overset{Me}{\underset{|}{C}}-C\equiv O^+ \ SbF_6^-$ (18)

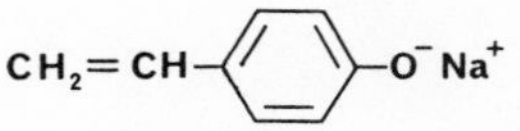

$CH_2=CH-C_6H_4-O^- \ Na^+$ (19)

$CH_2=CH-C_6H_4-CH_2O^- \ Na^+$ (20)

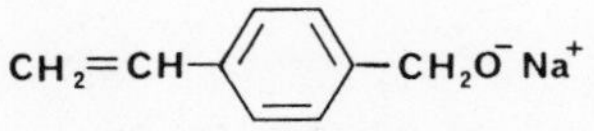

Macromers can also be prepared from oxiranes but only if polymerization is carried out by the activated monomer mechanism as already indicated. For such monomers, methacrylic end-groups, for example, can be introduced using 2-hydroxyethyl methacrylate as the functionalized initiator.[91] To prepare similarly functionalized macromers from aziridines, termination with methacrylic acid has been employed.[91]

Macromers have been prepared also from oxazolines.[92-94] The polymerization of 2-phenyl-2-oxazoline using p-iodomethylstyrene to form a functionalized initiator gives macromers with styrenic end-groups whilst the termination of similar polymerizations with N,N-dimethylaminopropyl methacrylamide gives oligomers with methacrylamido end-groups.[92] End-capping esterifications of polymerizations of 2-alkyl oxazolines with acrylic or methacrylic acid give oligomers with acrylate or methacrylate end-groups.[93,94]

Some stereoregular polysaccharide macromers with terminal acrylic or methacrylic groups have been produced recently by the cationic ring-opening polymerization of a silylated 1,4-anhydroribopyranose using acryloyl or methacryloyl chloride with silver hexafluorophosphate, or a similar silver salt, as the functionalized initiating system.[95]

The living polymerization of vinyl ethers initiated by HI/I_2 can also be used to prepare macromers as well as to prepare telechelics. For example, oligovinyl ethers with methacrylic end-functionality can be produced by initiation in the presence of the functionalized vinyl ether, (21).[96] Vinyl ether end-groups have been quantitatively introduced into similar oligomers by end-capping with the sodium salt of diethyl 2-vinyloxyethylmalonate.[97]

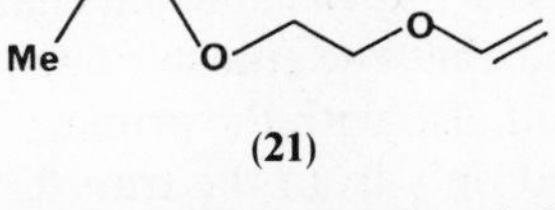

(21)

The preparation of macromers from certain olefins, such as isobutene, can be accomplished by variations of the inifer method outlined above. Thus, the use of the inifers (**22**) and (**23**) with BCl_3 to polymerize isobutene produces macromers with styrenic and norbornenyl end-groups respectively.[98,99]

(**22**)   (**23**)

6.2.4 By radical polymerization

6.2.4.1 General considerations As with the other mechanisms of addition polymerization, terminal functional groups can be introduced into polymers prepared by radical polymerization by one of two methods (or by a combination of them), viz. during the initiation step and/or during termination.

Radical methods, unfortunately, suffer from a number of disadvantages. In particular, if the introduction of functional groups is being attempted during initiation, either by using an appropriately functionalized radical initiator or an initiator which gives a radical of the appropriate functionality on decomposition (e.g. an hydroxyl radical), then it is important that chain termination takes place exclusively by combination of radicals if the resulting polymer chains are each to contain two end-groups of the required type and thus be truly telechelic. The requirement for termination exclusively by combination is probably not met by any monomers although styrene (and its derivatives), acrylates (but not methacrylates) and acrylonitrile are among monomers for which termination is predominantly by this mechanism. However, whether or not termination is normally by bimolecular combination of macroradicals is likely to be less important given the conditions under which the preparation of telechelics using functionalized initiators is usually attempted since, to get low molecular weight products, high concentrations of initiator are necessary and under these conditions much of the termination may involve primary radicals. Here too, though, it is important that the primary radical combines with the growing macroradical rather than undergo disproportionation. It is also a necessary condition of the 'initiator' method that the initiating radical adds to the monomer during initiation and does not initiate by hydrogen abstraction.

An alternative method by which to indroduce terminal functional groups into radical polymers makes use of efficient chain transfer agents. It is a requirement of this method that the transfer agent reacts not only with the growing polymer radical but also with the primary radical if the polymer is to contain end-groups derived only from the transfer agent. If this condition is not met then functionalized chain transfer agents can be used in concert with functionalized initiators.

As indicated in earlier sections, living anionic, group transfer and cationic polymerizations can be used to produce polymers, and therefore also oligomers, with very narrow (Poisson) distributions of molecular weight, i.e. with polydispersities approaching 1. With radical polymerization, however, the non-instantaneous nature of the initiation process and the random nature of chain termination combine to produce materials often with much higher polydispersities. For high molecular weight polymers produced by radical polymerization the lowest observed polydispersity is 1.5, i.e. when termination is by combination. When termination is by disproportionation or by transfer, the expected polydispersity is 2 (the 'most probable' distribution). At high conversions, polydispersities may be considerably higher even than 2 through the occurrence of transfer to polymer, through restriction of termination by the gel effect, or through the premature consumption of initiator. However, in practice the polydispersities of oligomers prepared by radical polymerization are often quite low; this is not as paradoxical as it might seem since when the average chain length is depressed by the use either of transfer agents or of high initiator concentrations the probability of forming long chains is much reduced. In fact as the chain length is reduced so the polydispersity tends towards 1, the value for a monoaddition product.

6.2.4.2 Telechelic oligomers

The initiator method. The most widely used group of initiators for the preparation of telechelics by radical means are azo compounds. Azobisisobutyronitrile, which is a general initiator of radical polymerization, can be used to produce oligomers with terminal nitrile groups. This initiator has been used to prepare oligostyrenes with terminal cyano functionalities very close to two, indicating that termination is indeed predominantly by combination. The average number of molecular weights of these oligomers ranged from 450 to 2100 and the polydispersities were around 1.5.[100] However, the use of the same initiator in high concentration with ethylene resulted in oligomers with significantly lower functionalities,[101] a result attributed in part to the occurrence of some termination by disproportionation. A variety of other functionalized azo-compounds containing hydroxyl, carboxyl and isocyanate groups have been used to produce telechelic oligomers from a variety of monomers, including from butadiene, isoprene, chloroprene, acrylonitrile, vinyl acetate, methyl acrylate and methyl methacrylate.[102] A recent example makes use of an azo-compound containing imidazole groups (**24**) together with a chain transfer agent to

(**24**)

prepare single-ended α-imidazole oligostyrenes and oligo(vinyl acetate)s[103,104] for use in synthesizing graft copolymers.

Peroxides appear on the whole to be less suitable than azo-compounds for the introduction of end-groups into oligomers. However, hydroxy-ended butadiene oligomers are produced successfully in a commercial process using hydrogen peroxide as initiator.[105] Such oligobutadienes have important uses as binders, for example, of solid rocket propellants. Some oligostyrenes with terminal chloromethyl or formal groups have been produced by using the appropriately p-substituted benzoyl peroxide as initiator.[106] However, because the benzoyloxy radical can dissociate to give phenyl radicals and carbon dioxide prior to initiation, the oligomers would undoubtedly have contained a mixture of end-groups of the types (25) and (26). Although these end-groups bear the same functional groups their reactivities could differ; also the benzoate ester type (25) is susceptible to hydrolysis. Other peroxides that have been used to produce telechelic oligomers include cyclohexanone peroxide and the peroxides of butane and pentane dioic acids.[107]

(25)   (26)

The use of dialkylperoxydicarbonates to prepare functionalized oligoethylenes is complicated by the effects of radical decomposition and transfer reactions so that the oligomers produced contain saturated and unsaturated alkyl end-groups in addition to the expected ester end-groups.[108] Peroxydicarbonates have also been used to prepare oligobutadienes with terminal carbonate groups; these groups could be converted to hydroxyl groups by hydrolysis.[109]

A novel initiator technique which might be used to introduce near-terminal functionality into an oligomer produced by radical polymerization involves the use of a comonomer which is extremely reactive towards the initiating radical but which fails to copolymerize with the principal monomer to a significant extent. Stilbene and its derivatives are possible examples; it has been shown recently that stilbenes are highly reactive towards the benzoyloxy radical and that in copolymerizations of stilbene with styrene initiated by benzoyl peroxide the bulk of the stilbene is located at the chain-ends adjacent to benzoate end-groups.[110]

The use of transfer agents, telogens and iniferters. The general mechanism for the radical polymerization of a monomer M in the presence of an efficient chain transfer agent XY is shown is Scheme 6.9. It is important if high yields of the telechelic oligomer X–(–M–)$_n$–Y are to be obtained that initiation takes place exclusively via reaction of primary radicals, R·, with the transfer agent and that chain transfer of the growing radical takes place much more

$$R\cdot + X-Y \longrightarrow R-Y + X\cdot$$

$$X\cdot + nM \longrightarrow X-(M)_n\cdot$$

$$X-(M)_n\cdot + X-Y \longrightarrow X-(M)_n-Y + X\cdot$$

Scheme 6.9

frequently than conventional bimolecular termination. The functionality of the oligomer and its degree of polymerization are thus both controlled by the reactivity of the transfer agent and its concentration relative to that of the monomer.

If the chain transfer agent is very reactive and the molecular weights of the products formed are consequently very low, the process is referred to as 'telomerization' and the monomer, transfer agent and product are referred to as a 'taxogen', 'telogen' and 'telomer' respectively. The use of telomers to produce telechelic oligomers has recently been well reviewed.[111] Particularly well-studied telogens are the carbon tetrahalides.[112-115] The use of carbon tetrachloride in polymerizations of ethylene and styrene,[112,113] for example, leads to oligomers having one Cl– end-group and one CCl_3– end-group. Although these end-groups are not inherently particularly reactive they can be converted to other groups that are, e.g. NH_2– and HOOC–.

Some chain transfer agents, principally organo-sulphur compounds, appear to be capable of acting also as initiators and as terminating agents. Such transfer agents have been termed 'iniferters'; their action is described in Section 1.2.2. An early example is 3-(3-acetyl phenyl diazothio)acetyl benzene (**27**). This compound when used to initiate the polymerization of butadiene produces oligomers with $MeCOC_6H_4$– and $MeCOC_6H_4S$– end-groups.[116] Disulphides such as tetraethyldithiuramdisulphide and benzoyl disulphide also function as iniferters and can be used to produce α,ω-bifunctional oligostyrenes and oligo(methyl methacrylate)s for example. The use of these and other similar sulphur-based iniferters has recently been reviewed.[117]

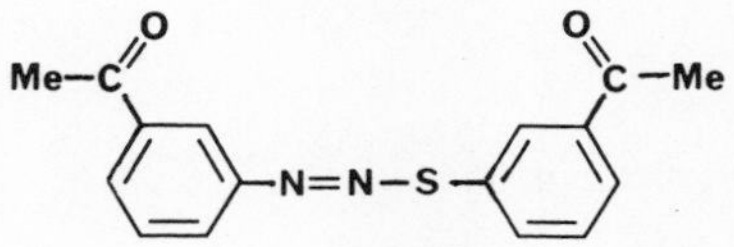

(**27**)

Other compounds which also probably function in much the same way as the disulphide iniferters are tetraphenylethanes. For example, (**28**) and (**29**) have been used to prepare oligomers of methyl methacrylate with diphenylphenoxy-methyl end-groups and of styrene with trimethylsiloxane end-groups respectively.[118-121]

An unusual example of the use of a functionalized transfer agent to produce

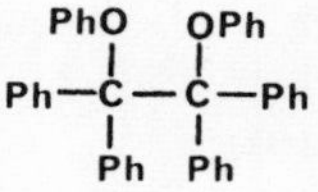

178   NEW METHODS OF POLYMER SYNTHESIS

$$\text{Ph}-\underset{\underset{\text{Ph}}{|}}{\overset{\overset{\text{PhO}}{|}}{C}}-\underset{\underset{\text{Ph}}{|}}{\overset{\overset{\text{OPh}}{|}}{C}}-\text{Ph} \qquad \text{Me}_3\text{Si}-\text{O}-\underset{\underset{\text{Ph}}{|}}{\overset{\overset{\text{Ph}}{|}}{C}}-\underset{\underset{\text{Ph}}{|}}{\overset{\overset{\text{Ph}}{|}}{C}}-\text{O}-\text{SiMe}_3$$

(28) 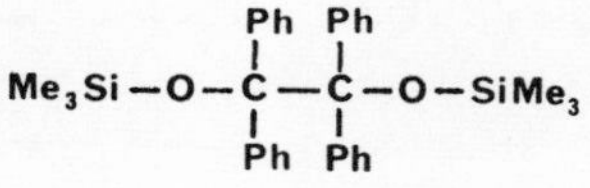   (29)

telechelic oligomers is the use of functionalized ketene acetals to produce oligostyrenes with terminal alkoxycarbonyl end-groups by an addition-elimination (Scheme 6.10).[122]

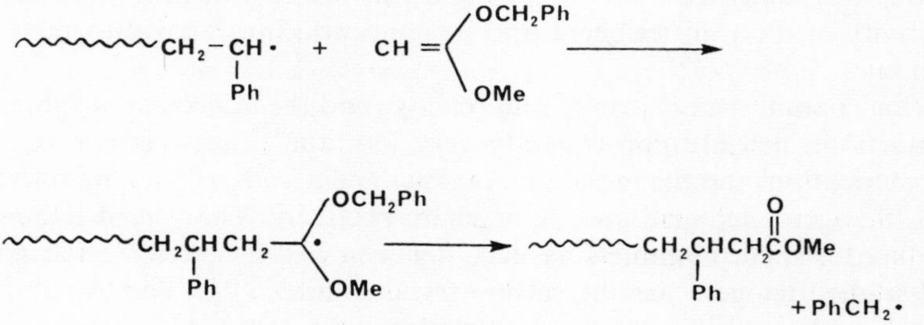

Scheme 6.10

6.2.4.3 *Macromers*   Macromers can be produced by further reactions with a suitably unsaturated reagent of terminally functionalized oligomers made by any of the above methods. However, a direct synthesis of some poly(methyl methacrylate) macromers has been achieved by catalysing the chain transfer of the growing radical to monomer with either cobalt porphyrin derivatives[123] or cobaloximes[124] (equation 6.7). The resulting macromers copolymerized well with ethyl acrylate.[125] Similar macromers were also prepared from methacrylonitrile.[125]

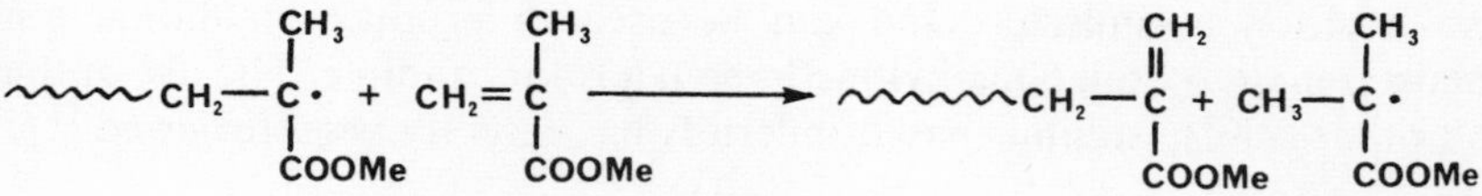

(6.7)

6.2.5 *By stepwise polymerization*

6.2.5.1 *General considerations*   Owing to the nature of the stepwise polymerization process, i.e. a process in which functional groups on monomers react progressively, either by condensation or rearrangement, to give higher molecular weight species carrying the same functional groups, a wide range of terminally functional oligomers can be made by this means.

However, oligomers made by stepwise polymerizations will suffer one of the disadvantages of those made by radical addition polymerization, viz. the molecular weight distribution will be rather broad. In general, $\bar{x}_n$, the number average degree of polymerization, will be given by $\bar{x}_n = 1/(1-p)$ where p is the extent of reaction (the fraction of functional groups reacted), whilst $\bar{x}_w$, the weight average degree of polymerization, will be given by $\bar{x}_w = (1+p)/(1-p)$. Therefore the polydispersity ($\bar{x}_w/\bar{x}_n$) will be equal to $1+p$ and will approach 2 as p tends to 1.

In the case of stepwise polymerizations involving the self-condensation or self-rearrangement of an α,ω-difunctional monomer with two different functional groups (e.g. an α,ω-amino acid) there are two strategies by which the molecular weight of the product can be limited and hence oligomers be obtained. The first of these is to limit the extent of conversion, p, and the second is to add a small amount of a monofunctional reagent as a chain stopper. In the presence of a chain stopper, $\bar{x}_n = (1+q)/(1-p+q)$ where q is the mole ratio of monofunctional reagent to difunctional reagent. The use of a chain stopper will, of course, produce only a monofunctional oligomer unless the stopper also carries a functional group which can be utilized in later reactions but which does not take part in the stepwise polymerization process itself.

If the stepwise polymerization involves the use of two difunctional reagents, e.g. an α,ω-diacid reacted with an α,ω-diamine, then there is also a third way of limiting molecular weight. This is to employ unequal amounts of the two components. In such a system $\bar{x}_n = (1+r)/(1+r-2rp)$ where r is the stoichiometric imbalance (the mole ratio of the minor component to the major component). (Note: In this instance p is defined as the extent of reaction of the minor component. The end-groups will be those of the major component.)

All of the above devices have been used to produce telechelic oligomers of controlled average molecular weight. The literature on this area is extensive and has recently been thoroughly reviewed.[5] Here, therefore, only a few representative examples are chosen, particularly those involving the synthesis of terminally functional oligomers based on aromatic monomers. Such oligomers are useful intermediates, for example, for the synthesis of hard blocks in thermoplastic elastomers and of easy-to-process thermally stable polymers.

6.2.5.2 *Telechelic oligomers* A familiar example of a stepwise polymerization carried out under non-stoichiometric conditions is the reaction of bisphenol A with an excess of epichlorohydrin in the presence of base to form the diglycidyl ether (**30**), an important intermediate for the synthesis of epoxy resins.

(**30**)

Reactions under non-stoichiometric conditions have been used also to prepare oligoimides from aromatic dianhydrides and aromatic diamines,[126] oligoquinolines from aromatic diamines and aromatic diketones,[127] oligo(ether sulphone)s from the sodium or potassium salt of bisphenol A with bis(4-chlorophenyl)sulphone[128,129] (see Chapter 1, equation 1.6) and hydroxy-terminated oligo(phenylene oxide)s from oxidative copolymerizations of mono- and di-functional phenols (equation 6.8).[130,131] A recent example is of the synthesis of aromatic oligomers from 1,4-diethynylbenzene and 1,3-benezenedithiol having either —SH or —C≡CH terminal groups depending upon the stoichiometric ratio of the monomers.[132]

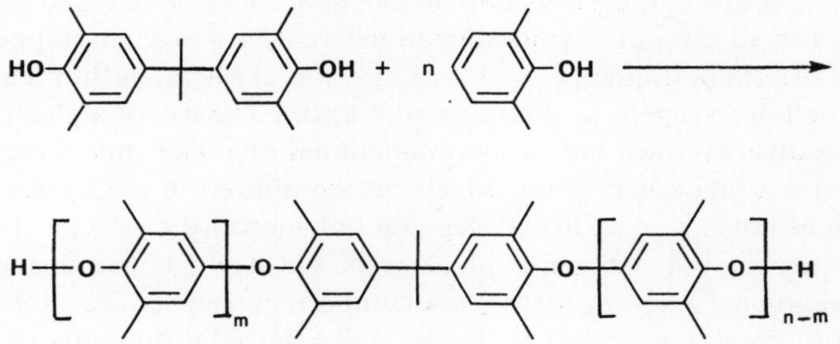

(6.8)

An ingenious example of the use of chain stoppers to control molecular weight and functionality involves the reaction of bisphenol A, in which one hydroxyl group has been blocked by prior reaction with trimethylchlorosilane, with phosgene to form an oligocarbonate (Scheme 6.11).[133]

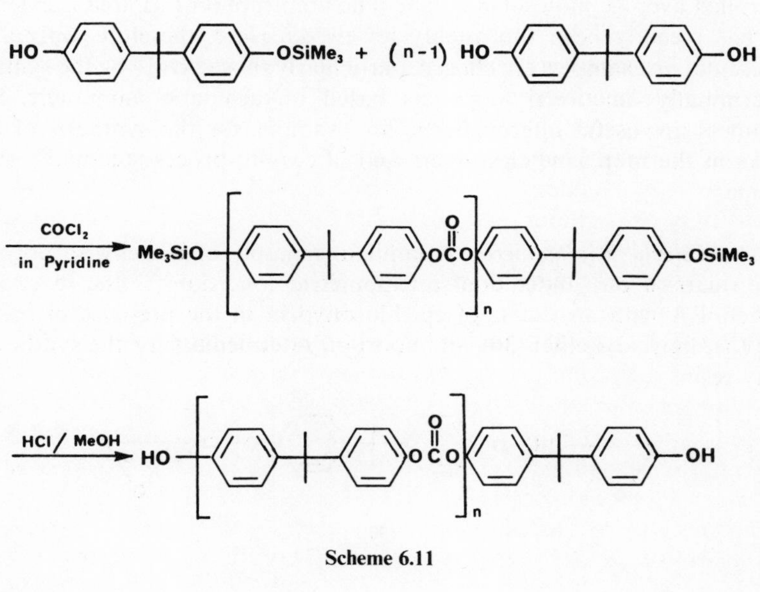

Scheme 6.11

6.2.5.3 Macromers
Macromers are most conveniently synthesized from telechelics that have been made by stepwise polymerization by a secondary reaction of the terminal functional groups on the telechelic with a reagent which introduces the necessary unsaturation. Some reactions of this type are considered later.

6.2.6 By controlled polymer chain scission

6.2.6.1 Telechelic oligomers
The controlled scission of polymer chains offers the potential for making telechelic oligomers from any high molecular weight polymer regardless of the mechanism by which it has been formed. The method also offers the possibility of high end-functionality. For example, to prepare an oligomer with an average molecular weight of 2000 from a polymer of average molecular weight 200 000 requires, on average, 99 scissions per chain. If each chain scission is carried out in such a way that useful terminal functional groups are introduced at that point (one at each side of the scission), 98 of the 100 oligomer chains produced on average from each polymer chain will be difunctional and only 2 will be monofunctional (assuming that the original polymer end-groups were not appropriately functionalized). The more scissions per chain that are carried out in order to obtain the oligomer then the higher will be the average functionality.

The disadvantage of the chain scission method is, assuming scission takes place at random along the polymer chain, that the distribution of oligomer chain lengths will tend towards the most probable distribution (polydispersity between 1 and 2 depending upon average chain length) as with oligomers made by radical and stepwise processes.

Scission of homopolymers. One of the most widely employed reactions for generating telechelic oligomers from homopolymers is oxidative cleavage. For example, poly(vinyl chloride)[134] and polybutadiene[135] have both been ozonized to give telechelic oligomers carrying –OH end-groups after reductive work-up of the ozonides. In both cases cleavage occurs at double bonds; in the case of poly(vinyl chloride) these double bonds are structural defects in the polymer, e.g. arising as a result of slight dehydrochlorination. An ingenious method of obtaining telechelic oligomers with terminal carboxylic acid groups from polyethylene is to selectively oxidize polyethylene single crystals at the amorphous chain folds using either ozone or fuming nitric acid.[136,137] The oligomers so formed have lengths corresponding to the thickness of the crystal lamellae. A similar strategy has been employed recently to prepare telechelic oligoisoprenes having very narrow polydispersities (1.06) by treating with ozone partly crystalline high cis-1,4-polyisoprenes deposited upon cooling solutions of the polymers in hexane to $-70\,°C$.[138]

Another type of reaction which has been used extensively to produce telechelic oligomers is hydrolysis. α-Hydroxy-ω-carboxy oligocaproamide, for example, has been made by hydrolysing Nylon-6 using sodium nitrite in acid solution.[139]

Scission of copolymers. A more elegant and more controllable way of producing telechelic oligomers by chain scission is to employ, as starting materials, polymers which contain deliberately introduced and randomly distributed 'weak links'. Early examples of this strategy involve the use of ozone to degrade copolymers of isobutene with a variety of dienes (equation 6.9).[140-142]

$$\sim\sim\sim CH_2-\underset{\underset{Me}{|}}{\overset{\overset{Me}{|}}{C}}-CH_2-\underset{}{\overset{\overset{R}{|}}{C}}=\underset{}{\overset{\overset{R}{|}}{C}}-CH_2-CH_2-\underset{\underset{Me}{|}}{\overset{\overset{Me}{|}}{C}}\sim\sim\sim \xrightarrow{O_3}$$

$$\sim\sim\sim CH_2-\underset{\underset{Me}{|}}{\overset{\overset{Me}{|}}{C}}-CH_2-C\overset{R}{\underset{O}{\diagdown\!\!\!\!\diagup}} \;+\; \overset{R}{\underset{O}{\diagdown\!\!\!\!\diagup}}C-CH_2-CH_2-\underset{\underset{Me}{|}}{\overset{\overset{Me}{|}}{C}}\sim\sim\sim$$

(6.9)

Recently, ozonolyses of high molecular weight (~200 000) copolymers of methyl methacrylate with small amounts of butadiene, isoprene and 2,3-dimethylbutadiene have been used to prepare telechelic oligomethyl methacrylates with terminal carboxylic acid and methyl ketone groups.[143] The functionalities of these oligomers were very close to 2 and their polydispersities ranged from 1.2 to 1.8 depending upon molecular weight. It has been demonstrated that for some copolymers containing double bonds, oxidative cleavage may be accomplished more cleanly using ruthenium tetroxide rather than ozone.[144]

Readily cleavable polymer chains can be made by copolymerizing vinyl monomers, such as styrene, ethylene, vinyl acetate and vinyl chloride, with 2-methylene-1,3-dioxepane which undergoes ring-opening polymerization to introduce in-chain ester links[145,146] (see Section 1.2.4). Subsequent alkaline hydrolysis of the ester links produces oligomers with α-hydroxy-ω-carboxy functionality. Degradation of such copolymers can also be carried out enzymatically; such a method has been used with copolymers of 2-methylene-1,3-dioxepane with ethylene. 2-Methylene dioxolane also can be copolymerized to introduce hydrolyzable ester links, although in this case there can be some incorporation of the monomer by polymerization through the exocyclic double bond without ring-opening.

Ring-opening copolymerization can also be applied to spiro-o-carbonates. Copolymers of spiro-o-carbonates with styrene and hyroxyethyl methacrylate when hydrolysed give oligomers with terminal hydroxyl groups.[147]

6.2.6.2 *Macromers*   Whilst there are as yet no reports of the direct synthesis of macromers by polymer chain scission, macromers could be made by appropriate further reactions of telechelics made by the chain scission process. Some typical reactions of this type are outlined below.

6.2.7 Modifications of end-groups of telechelic oligomers

6.2.7.1 To make other telechelic oligomers Modifications of the end-groups of telechelic oligomers may be a necessary second stage following their syntheses for two reasons: either because the primary end-groups are not of the correct type to take part in the further reactions envisaged in converting the oligomers into some high molecular weight product, or because the end-groups are of too low a reactivity to react to give high molecular weight materials at a convenient temperature. The types of reaction which have been employed to modify oligomer end-groups are legion; in fact almost any organic reaction could be envisaged as being appropriate to certain types of end-group modification. The choice of reaction in a particular case will be largely dictated by the nature of the existing end-group and of the end-group which it is desired to achieve, but also it may be limited by the structure of the oligomer repeat unit. As an example, it would obviously be unwise to attempt to modify the end-groups of an oligomer by alkaline hydrolysis if the oligomer contained ester groups, whether in the backbone or as side-groups. Various reactions which have been used to modify end-groups have been well reviewed[5] and therefore only representative and recent examples are given here.

The most often cited modifications are to telechelic oligomers bearing hydroxyl end-groups. Such end-groups can be readily converted, for example, to isocyanate groups by reaction with an excess of a diisocyanate; such isocyanate-ended oligomers lend themselves to rapid reactive processing applications.[148] Hydroxyl groups can also be converted to chloroformate by reaction with phosgene;[149] to nitro by reaction with p-nitrobenzoylchloride;[150] to epoxy by reaction with epichlorohydrin[151-153] and to carboxylic acid either by direct oxidation[154] or, in cases where the repeat unit might also be oxidized, by reaction with a diacid dichloride followed by hydrolysis.[155]

Carboxylic acid end-groups can often be converted to hydroxyl by direct reduction. For example, α,ω-carboxy oligo(methyl methacrylate)s have been converted to the corresponding hydroxyl-ended materials by reduction with borane in THF.[143] Carboxylic acid groups can also be converted to acid chloride[156] by reaction with phosphorus pentachloride or thionyl chloride, and to isocyanate groups in a two-stage process involving reaction first with ethylchloroformate and then with sodium azide[157] (Scheme 6.12). Aromatic acid anhydride end-groups in telechelic precursors to polyimides are commonly reacted with a variety of appropriately substituted aromatic amines in order to introduce cross-linkable functionality for use in subsequent processing.[126]

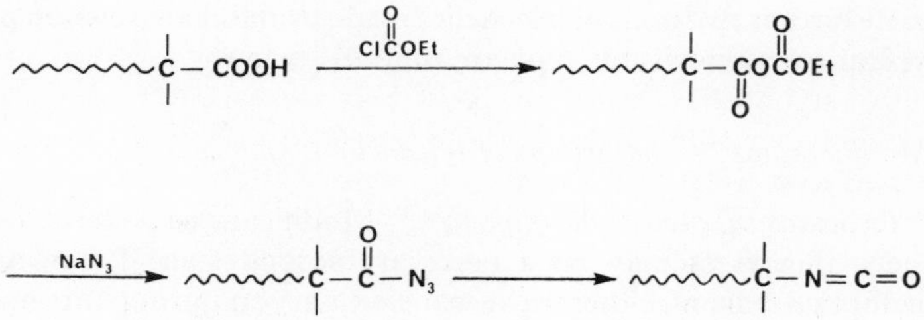

Scheme 6.12

Amine-ended oligomers can be prepared by reduction of either nitrile- or nitro-ended materials. Thus oligostyrenes with terminal nitrile groups made by radical polymerization using azoisobutyronitrile as initiator have been so converted[157,158] as have ethylene oxide oligomers with terminal nitrobenzyl groups.[159] Polyisobutenes with terminal phenyl groups have been given amine-functionality by first nitrating with HNO_3/H_2SO_4 and then reducing the resultant, mainly *para*, nitro-groups with stannous chloride.[160]

6.2.7.2 *To make macromers* Chemical modifications of the end-groups of telechelic oligomers are particularly important for the synthesis of macromers which cannot be made directly by oligomerizations involving the use of functionalized initiators or terminating agents. Macromers with acrylic or methacrylic end-groups have commonly been synthesized from appropriately end-functionalized oligomers by reaction with acrylic or methacrylic acid, with their acid chlorides, or with some other derivative, such as 2-hydroxyethyl methacrylate, glycidyl methacrylate or 2-isocyanatoethyl methacrylate. For example, ω-acryloyl oligo(vinylidene chloride) has been synthesized by esterifying hydroxy-ended oligomers with acrylic acid[161] and acrylate and methacrylate derivatives of various proteins have been made by reacting the terminal amino groups with the corresponding acid chlorides.[162] Similar strategies have been used to prepare ω-acrylamido oligo(triphenylmethyl methacrylate)[163] and ω-acrylamido and ω-acryloyl oligoisobutenes.[151,160,164]

2-Hydroxyethyl methacrylate has been reacted with an ω-isocyanato oligourethane to give an ω-methacryloyl urethane macromer[165] and glycidyl methacrylate has been used to prepare hydrophilically ended ω-hydroxypropylmethacryloyl macromers from ω-carboxy oligostyrenes[166,167] and acrylates.[168,169] 2-Isocyanatoethyl methacrylate reacts smoothly with ω-hydroxy oligo(ethylene oxide)s to form the ω-methacryloyl macromers.[170]

Macromers with ω-(*p*-vinylphenyl) or ω-(*p*-vinylbenzyl) functionality, and therefore which will have styrene-like reactivity in polymerization, can be synthesized by reacting telechelic oligomers with suitable *p*-substituted styrene derivatives. Commonly exploited *p*-substituents include $-CH_2Cl$, $-NH_2, -CH_2NH_2$ and $-MgCl$. For example, ω-(*p*-vinylbenzyl) oligo(ethylene oxide)s and oligo(propylene oxide)s have been prepared by etherifying the corresponding ω-hydroxy oligomers with p-chloromethylstyrene.[171,172] p-Chloromethyl styrene has been used also to prepare similar macromers from aromatic oligo(ether sulphone)s.[173,174] p-Vinylphenyl magnesium chloride has been used to convert ω-chloro oligodimethylsiloxanes to ω-vinylphenyl silicone macromers.[175] p-Aminomethyl styrene has been reacted with various oligopeptides[176] and oligosaccharides[177] to convert them to macromers. Polymerizable styrenic and other groups have also been attached to lignin residues.[178]

Polymerizable fumarate-ended oligo(ethylene oxide)s have been synthesized by reacting ω-hydroxy oligomers with maleic anhydride or methyl fumarate in the presence of 4-dimethylaminopyridine.[179]

Some novel condensation-type macromers with terminal anhydride functionality based on methyl methacrylate have recently been made by reacting mellitic trianhydride with ω-amino or ω-hydroxy oligo(methyl methacrylate). These macromers could be condensed with bisphenol-A to produce aromatic polyester-poly(methyl methacrylate) graft copolymers[180] or be reacted with p-aminobenzoic acid and m-phenylenediamine to give aromatic polyamide-poly(methyl methacrylate) graft copolymers.[181]

6.3 Reactions and reactivity of telechelic oligomers and macromers

6.3.1 Telechelic oligomers

The major use of telechelic oligomers is for the synthesis of block copolymers. Typically, telechelic oligomers can be reacted in ways similar to those used with difunctional low molecular weight monomers. i.e. by stepwise reactions involving either condensation or rearrangement. Basically, three strategies can be used: (i) the two telechelic oligomers can be directly reacted either in the melt or in solution (equation 6.10); (ii) they can be linked with a difunctional compound of low molecular weight (equation 6.11); or (iii) one telechelic oligomer can be reacted with the precursors of another (equation 6.12). The choice of strategy will be determined to a large extent by the end-groups present and whether or not they are mutually reactive. The formation of block copolymers by these routes has been extensively reviewed[5] and it will suffice to

$$X\sim\sim\sim X + Y\sim\sim\sim Y \longrightarrow \sim\sim X'Y'\sim\sim\sim Y'X'\sim\sim \quad (6.10)$$

$$X\sim\sim\sim X + A-B + Y\sim\sim\sim Y \longrightarrow \sim\sim X'A'-B'Y'\sim\sim \quad (6.11)$$

$$X\sim\sim\sim X + A-A + B-B \longrightarrow \sim\sim X'(A'-B')_n X'\sim\sim X'(A'-B')_m X'\sim \quad (6.12)$$

give here just a few typical examples and to consider some characteristic features and some of the problems that may arise.

Strategy (i) above has been used, for example, to prepare polysiloxanes containing alternate aromatic and aliphatic siloxane sequences by the co-condensation of α,ω-dihydroxy aliphatic and aromatic oligosiloxanes,[182] to prepare a siloxane-aromatic sulphone block copolymer from an α,ω-diepoxy oligosiloxane and an α,ω-diphenolic oligosulphone,[183] to prepare a poly(amide-ether) block copolymer by condensing an α,ω-dicarboxy oligoamide with an α,ω-hydroxy oligoether,[184] and to synthesize a polymer with butadiene and sulphone blocks from an α,ω-dichlorocarboxy oligobutadiene and an α,ω-diphenoxy oligoethersulphone.[152]

Examples of strategy (ii) are the co-condensations of α,ω-diphenolic oligo(arylene ether)s and oligo(arylene sulphone)s with α,ω-diphenolic oligocarbonates using phosgene as the linking agent.[185,186] Strategy (iii) has been employed to make a copolymer containing oligodimethylsiloxane and oligo(bisphenol A polycarbonate) blocks; the polycarbonate was synthesized in the presence of an α,ω-diphenolic oligodimethylsiloxane.[187]

The kinetics of stepwise reactions involving telechelic oligomers should in theory be the same as those for similar reactions involving small molecules, e.g. first order in each reactant and second order overall in the case of strategy (i). However, in practice deviations from expected behaviour often occur. Sometimes these may be as a result of side-reactions such as cyclization[186,188] or elimination[189] but more commonly as a result of reactions between end-groups and groups within the main chains leading to branching and cross-linking. For example, telechelic oligomers with isocyanate, chloroformate and anhydride end-groups have all been shown to react with in-chain amide groups as well as with the terminal amino groups of oligoamides.[190,191]

Additionally, deviations from expected behaviour may be as a result of purely physical effects. The two most important of these are likely to be heterogeneity of the reaction system and high viscosity, both of which will be most severe for reactions carried out in the bulk. Generally, pairs of oligomers will be incompatible and in the bulk, initially at least, reaction may take place at an interface. Only when sufficient block copolymer has been formed to act as a compatibilizer will homogeneity be achieved. Initial heterogeneity was thought to be the reason, for example, why reactions between α,ω-dicarboxy oligoamides and α,ω-dihydroxy oligo(ethylene oxide)s showed deviations from overall second order behaviour at conversions below 75%.[184] High viscosities will lead to diffusion control of reactions and will thus lead to lower than expected rates. Carrying out reactions between telechelic oligomers in solution provides a way of avoiding both of the effects mentioned above. However, it must be remembered that the use of a solvent will lead to lower rates of reaction by reducing the concentrations of the reactive end-groups; these concentrations are already low (relative to the concentrations of nonreactive material) in telechelic oligomers.

6.3.2 Macromers

The principal use for macromers is for the synthesis of graft copolymers. Typically, these syntheses involve the radical copolymerization of a macromer with terminal unsaturation with a low molecular weight monomer. Since it is usually desired to introduce only occasional side-chains, these copolymerizations are generally carried out with a large excess of the low molecular weight monomer. Under such conditions, the kinetics of copolymerization can be considerably simplified since to a first approximation is is only necessary to consider two propagation steps: those involving addition of the small monomer, M_1, and the macromer, M_2, to a radical with a terminal M_1 unit (Scheme 6.13). Under such conditions, the instantaneous copolymer composition equation (see Chapter 1, equation 1.8) reduces to that shown in equation 6.13, which on integration yields equation 6.14.

$$\sim\!\!\sim\!\!\sim M_1\cdot \;+\; M_1 \;\longrightarrow\; \sim\!\!\sim\!\!\sim M_1\cdot \qquad k_{11}$$

$$\sim\!\!\sim\!\!\sim M_1\cdot \;+\; M_2 \;\longrightarrow\; \sim\!\!\sim\!\!\sim M_2\cdot \qquad k_{12}$$

Scheme 6.13

$$\frac{m_1}{m_2} = \frac{d[M_1]}{d[M_2]} = r_1 \frac{[M_1]}{[M_2]} \tag{6.13}$$

$$\log \frac{[M_1]_t}{[M_1]_o} = r_1 \log \frac{[M_2]_t}{[M_2]_o} \tag{6.14}$$

In theory, in order to determine the reactivity of the macromonomer relative to that of the small monomer ($k_{12}/k_{11} = 1/r_1$) it should be sufficient just to determine the composition of the copolymer at a single time t ($[M_1]_t/[M_2]_t$) for a feed of initial known composition ($[M_1]_0/[M_2]_0$). However, it has been shown that induction periods in the copolymerization, such as have been observed in copolymerizations of methyl and butyl methacrylate with ω-vinylbenzylether oligo(2,6-dimethyl-1,4-phenylene oxide)s, can render this method unreliable and that it is preferable to measure copolymer compositions at various times for the given initial feed and to plot a graph of:

$$\log([M_1]_t/[M_1]_0) \text{ vs. } \log([M_2]_t/[M_2]_0),$$

based on equation 6.14.[192] This graph should be a straight line with a slope equal to r_1 but will pass through the origin only if there is no induction period. Using this latter method, it has been shown that the reactivities of the above mentioned macromers of different chain lengths (ranging from 1000 to 27 000) are initially lower than those of low molecular weight model compounds of similar structure but increase with chain length until the molecular weights reach values (5000–7000) which give phase-separated graft copolymers; thereafter the reactivities decrease again. Low initial reactivities can be ascribed to steric shielding of the macromer double bonds.[193-197] The subsequent apparent increases in reactivity may be due to preferential solvation of the growing chains with macromer leading to local feed compositions which are richer in macromer than the average value for the system as a whole. Beyond a certain macromer molecular weight, however, increasing immiscibility of the macromer and the growing chain could lead to local depletion of macromer and a reversal of the trend observed at lower molecular weights.

Steric hindrance in polymerization of macromers is demonstrated also by the relative difficulty of homopolymerizing macromers to produce high molecular weight polymers,[198,199] by apparent changes in macromer reactivity with concentration[200] and by high monomer exponents in polymerization rate equations.[201]

A feature of graft copolymers made by the radical macromer technique can be the relatively broad distributions of chemical composition. Theoretical calculations show that this distribution should be narrower for grafts made with monodisperse macromers than for those made with polydisperse macromers, and that these distributions should become narrower with increasing degree of polymerization.[202] Fractionation by chemical composition of copolymers of methyl methacrylate with an ω-methyl methacryloyl oligodimethylsiloxane produced fractions differing in methyl methacrylate content by up to 24 wt% in good agreement with theoretical predictions.[203]

A further complicating feature of macromer polymerization can be chain transfer to macromer-derived side-chains which leads to steadily increasing polymer molecular weight as conversion increases. In copolymerizations of butyl acrylate with ω-styrenic oligo(ethylene oxide)s the transfer was shown to involve chiefly benzylic hydrogens adjacent to the styryl units from the macromer.[204,205]

Some macromers are also polymerizable by non-radical mechanisms. In group transfer polymerization, the reactivity of an ω-methacryloyl oligostyrene was apparently much higher in the presence of methyl methacrylate comonomer, illustrating once again the effects of steric hindrance on polymerization.[198] In anionic polymerization of ω-styryl oligostyrenes also, effects of steric hindrance are manifested in reactivity data.[206] As expected, however, graft copolymers produced by anionic copolymerizations of macromers have narrower distributions of chemical composition than similar copolymers produced by radical copolymerization.[207]

6.4 Uses of terminally functionalized oligomers

6.4.1 *Telechelic oligomers*

As already indicated, the major use of telechelic oligomers is in the synthesis of block copolymers. Such copolymers are becoming increasingly important as the major constituents of structural thermoplastics and of thermoplastic elastomers. The former application has attracted particular interest because of the possibility of combining the chemical process of converting the oligomers to high molecular weight polymer (chain extension and often also cross-linking) with the physical operation of moulding the consequent material to the desired shape. Reaction injection moulding (RIM), as the combined operation is termed, offers a number of advantages over conventional moulding operations such as injection moulding and extrusion.[148] Firstly, it is less consumptive of energy than conventional moulding processes since it does not require the melting of a preformed thermoplastic; secondly, the viscosities of the oligomers used in reactive processing are usually much lower than those of polymer melts making processing easier; and, thirdly, processing can be carried out at relatively low temperatures thus avoiding the risk of degrading the material.

However, RIM brings with it some special requirements, not least of which is the need for chain extension and cross-linking reactions which are rapid enough for the moulding operation to be completed in a reasonable time but not so rapid that the ingredients cannot be mixed satisfactorily prior to reaction. Usually also reactions are required which proceed without the evolution of volatiles which could lead to voids in the final moulded article. The exception to this latter requirement is when the final material is to be made in an expanded form, but here it is more usual deliberately to add an agent which liberates a gas at the temperature of reaction (an azo-compound or a low boiling hydro- or halocarbon) rather than to rely upon the adventitious liberation of volatiles. Mechanical strength in the final moulded article can be due to chemical cross-linking which leads to the production of a thermoset, but more often it arises mainly as a consequence of phase separation within the material which leads in turn to a type of physical cross-linking.

The vast majority of RIM applications utilize reactions of diisocyanates with α,ω-dihydroxy oligomers to produce polyurethanes (Scheme 6.14). The major oligomers used are oligo(propylene oxide)s capped with ethylene oxide to provide terminal primary hydroxyl groups and ranging in molecular weight from c. 2000 to c. 7000. Low molecular weight chain extenders compatible with the oligomers such as 1,4-butanediol and ethylene glycol are used. The most commonly used diisocyanate is 4,4'-diphenylmethane diisocyanate (MDI).

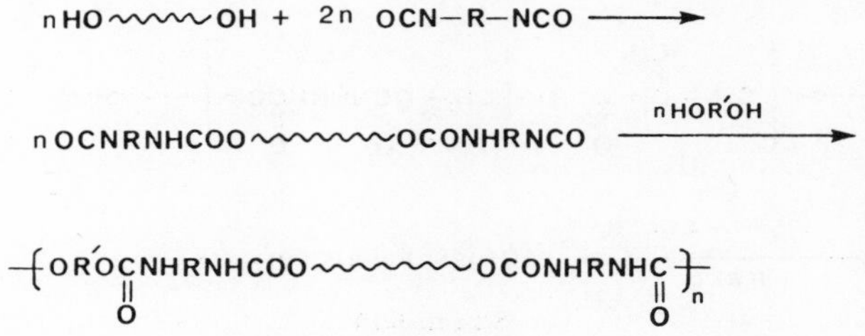

$$n\,HO\!\sim\!\!\sim\!\!\sim\!OH + 2n\,OCN-R-NCO \longrightarrow$$

$$n\,OCNRNHCOO\!\sim\!\!\sim\!\!\sim\!OCONHRNCO \xrightarrow{n\,HOR'OH}$$

$$-\!\!\left[OR'OCNHRNHCOO\!\sim\!\!\sim\!\!\sim\!OCONHRNHC\right]_n\!-$$
$$\|\|$$
$$OO$$

Scheme 6.14

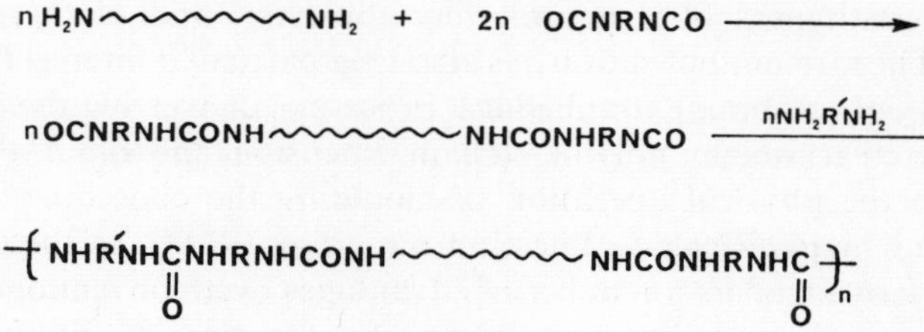

$$nH_2N\sim\sim\sim NH_2 + 2n\ OCNRNCO \longrightarrow$$

$$nOCNRNHCONH\sim\sim\sim NHCONHRNCO \xrightarrow{nNH_2R'NH_2}$$

$$-\!\!\!\!-\!\!\{NHR'NHCNHRNHCONH\sim\sim\sim NHCONHRNHC\}_n-\!\!\!\!-$$
$$\|\|$$
$$OO$$

Scheme 6.15

Polyureas also are used in RIM and are based upon reactions of diisocyanates with amine-terminated oligoethers (Scheme 6.15); they are chain extended with hindered diamines such as 2,4-diaminotoluene (non-hindered diamines react too quickly to be useful). Also increasingly used in RIM, especially for making automotive components, are hybrid urethane/urea systems.[208]

An interesting type of telechelic is one in which the terminal groups are ionic, e.g. carboxylate-terminated oligobutadienes. Such telechelics are a type of 'ionomer' and the ionic groups can form aggregates both in the solid state and in solution. In the solid state aggregation results in a form of physical cross-linking and, not surprisingly, ionomers have attracted interest as possible components of polymer blends.[209] Aggregation in solution leads to unusually high solution viscosities and as a consequence ionomers show potential as viscosifiers.[210] Other uses for telechelic oligomers include impact modifiers in thermosetting resins and in thermoplastics, as viscosity modifiers, as plasticizers and as emulsifiers.

6.4.2 Macromers

Macromers too are increasingly being investigated as components in RIM systems. An interesting commercial material (MODAR) is based upon reaction of hydroxyethyl methacrylate with an α,ω-diisocyanato oligoether

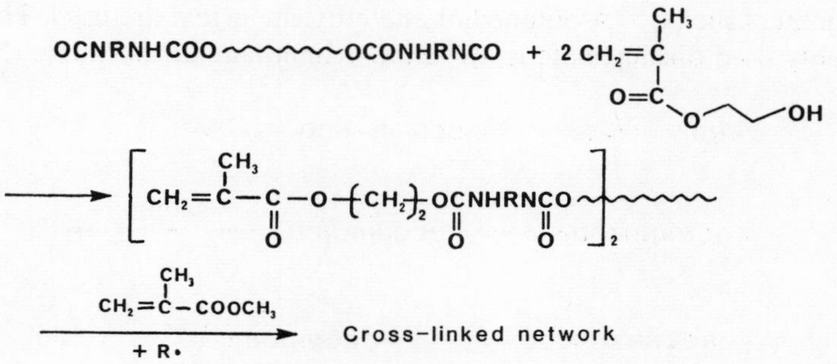

$$OCNRNHCOO\sim\sim\sim OCONHRNCO + 2\ CH_2=\overset{CH_3}{\underset{\underset{O\sim OH}{O=C}}{C}}$$

$$\longrightarrow \left[CH_2=\overset{CH_3}{\underset{}{C}}-\overset{}{\underset{\|}{C}}-O-(CH_2)_2OCNHRNCO\sim\sim\sim\right]_2$$

$$\xrightarrow[+R\cdot]{CH_2=\overset{CH_3}{\underset{}{C}}-COOCH_3} \text{Cross-linked network}$$

Scheme 6.16

to form a 'urethane methacrylate'. This urethane methacrylate is then copolymerized in a mould with methyl methacrylate monomer and a free radical initiator to give a cross-linked thermoset product, particularly suitable when reinforced with glass fibre, for automotive parts, pipe and bin linings, and casings for electrical equipment (Scheme 6.16).[211]

Other recently described applications of macromers include the use of oligodimethylsiloxane macromers copolymerized with 7,7,8,8-tetracyanoquinodimethane to make oxygen selective membranes[212], ω-methacryloyl oligo(perfluoroethylene oxide) copolymerized with methyl methacrylate to give a material for contact lenses,[213] acrylic and vinyl ether terminated oligoether macromers used to make solid electrolytes with potential battery applications,[214] hydrogels with potential drug-release applications based on polyether macromers copolymerized with hydroxyethyl methacrylate,[215] styrene-terminated macromers copolymerized with acrylic monomers to give pressure-sensitive adhesives,[216] 2-hydroxyethyl methacrylate macromers grafted onto polyethylene surfaces to provide an antithrombogenic surface for tubes, catheters, artificial blood vessels and transplants,[217] methacrylate terminated oligostyrene macromers used in making a polymer-modified concrete[218] and Nylon-6 with a water repellent surface made by incorporating a condensation-type macromer with a perfluoroalkyl side-chain.[219]

References

1. Uraneck, C.A., Hseih, H. and Buck. O.G. *J. Polym. Sci.* **46**, 535 (1960).
2. Milcovich, R. and Chiang, M.T. US Patent, 3,786,116 (1974).
3. Heitz, W. *Angew. Makromol. Chem.* **145–146**, 37 (1986).
4. Goethals, E.J. (Ed.), *Telechelic Polymers: Synthesis and Applications*, CRC Press, Boca Raton (1987).
5. Nguyen, H.A. and Marechal, E. *J. Macromol. Sci., Rev. Macromol. Chem. Phys.* **C28**, 187 (1988).
6. Percec, V., Pugh, C., Nuyken, O. and Pask, S.D. *Comprehensive Polymer Science* (Eds. Allen, G. and Bevington, J.C.) Pergamon Press, Oxford, Vol. 6, Ch. 9 (1989).
7. Nuyken, O. and Pask, S.D. *Encyclopedia of Polymer Science and Engineering* (Ed. Kroschwitz, J.I.) Wiley, New York, (2nd edn.), Vol. 16, pp. 494–532 (1989).
8. Richards, D.H. and Szwarc, M. *Trans. Faraday Soc.* **55**, 1644 (1959).
9. Hayashi, K. and Marvel, C.S. *J. Polym. Sci., Part A-1* **2**, 2571 (1964).
10. Morton, M., Fetters, L.J., Inomater, J., Rubio, D.C. and Young, R.N. *Rubber Chem. Technol.* **49**, 303 (1976).
11. Richards, D.H. in *Developments in Polymerization 1* (Ed. Haward, R.N.), Applied Science, London p. 1 (1979).
12. Hirao, A., Hattori, I., Sasagawa, T., Yamaguchi, K., Nakahama, S. and Yamazaki, N. *Makromol. Chem., Rapid Commun.* **3**, 59 (1982).
13. Richards, D.H., Service, D.M. and Stewart, M.J. *Brit. Polym. J.* **16**, 117 (1984).
14. Morton, M. and Mikesell, S.L. *J. Macromol. Sci., Chem.* **A7**, 1391 (1973).
15. Burgess, F.J. and Richards, D.H. *Polymer* **17**, 1020 (1976).
16. Luston, J. and Vass, S. *Adv. Polym. Sci.* **56**, 91 (1984).
17. Burgess, F.J., Cunliffe, A.V., Richards, D.H. and Sherrington, D.C. *J. Polym. Sci., Polym. Lett. Ed.* **14**, 471 (1976).
18. Burgess, F.J., Cunliffe, A.V., MacCallum, J.R. and Richards, D.H. *Polymer* **18**, 719 (1977).

19. Davis, A., Richards, D.H. and Scilly, N.F. *Makromol. Chem.*, **152**, 133 (1972).
20. Burgess, F.J., Cunliffe, A.V., MacCallum, J.R. and Richards, D.H. *Polymer* **18**, 726 (1977).
21. Cohen, P., Abadie, M.J.M., Schue, F. and Richards, D.H. *Polymer* **22**, 1316 (1981).
22. Schultz, D.N., Halasa, A.F. and Oberster, A.E. *J. Polym. Sci., Polym. Chem. Ed.* **12**, 153 (1974).
23. Schultz, D.N. and Halasa, A.F. *J. Polym. Sci., Polym. Chem. Ed.* **15**, 2401 (1977).
24. Price, C.C. and Carmelite, D.D. *J. Amer. Chem. Soc.* **88**, 4039 (1986).
25. Kazanskii, K.S., Solovyanov, A.A. and Entelis, S.E. *Eur. Polym. J.* **7**, 1421 (1971).
26. Cabasso, I. and Zhilka, A. *J. Macromol. Sci., Chem.* **A8**, 1313 (1974).
27. Boileau, S., Champetier, G. and Sigwalt, P. *Makromol. Chem.* **69**, 180 (1963).
28. Bergwerf, W., Wagner, W. and Marius, K. (Shell Int. Research), British Patent, 1,133,293 (1968).
29. King, C. (E.I. DuPont de Nemours), US Patent, 3,418,393 (1968).
30. Wilson, D.R. and Beaman, R.G. *J. Polym. Sci., Part A-1* **8**, 2161 (1970).
31. Hamitou, A., Ouhadi, T., Jerome, R. and Teyssie, P. *J. Polym. Sci., Polym. Chem. Ed.* **15**, 865 (1977).
32. Aida, T., Sanulki, K. and Inoue, S. *Macromolecules* **18**, 1049 (1985).
33. Sormani, P.M., Minton, R.J. and McGrath, J.E. *ACS Symp. Ser.* **286**, 147 (1985).
34. Yilgor, I., Riffle, J.S. and McGrath, J.E. *ACS Symp. Ser.* **282** 161 (1985).
35. Steckle, Jr., W.P., Yilgor, E., Riffle, J.S., Spinu, M., Yilgor, I. and Ward, R.S. *Polym. Prepr.* **28**(1), 254 (1987).
36. Chiang, M.T. and Milkovich, R. (CPC International Inc.), Ger. Offen., 2,208,340 (1972).
37. Schulz, G.O. and Milkovich, R. *J. Appl. Polym. Sci.* **27**, 4773 (1982).
38. Masson, P., Franta, E. and Rempp, P. *Markromol. Chem., Rapid Commun.* **3**, 499 (1982).
39. Chaumont, P., Herz, J. and Rempp, P. *Eur. Polym. J.* **15**, 459 (1979).
40. Asami, R., Takaki, M. and Hanahata, H. *Macromolecules* **16**, 628 (1983).
41. Norton, R.L. and McCarthy, T.J. *Polym. Prepr.* **28**(1), 174 (1987).
42. Rao, P.R., Masson, P., Lutz, P., Beinert, G. and Rempp, P. *Polym. Bull.* **11**, 115 (1984).
43. Takaki, M., Asami, R., Tanaka, S., Hayashi, H. and Hogen-Esch, T.E. *Macromolecules* **19**, 2900 (1986).
44. Severini, F., Pegoraro, M. and Saija, L. *Angew. Makromol. Chem.* **133**, 111 (1985).
45. Masson, P., Beinert, G., Franta, E. and Rempp, P. *Polym. Bull.* **7**, 17 (1982).
46. Sigwalt, P. *Angew. Makromol. Chem.*, **94**, 1101 (1981).
47. Hamaide, T., Revillon, A. and Guyot, A. *Eur. Polym. J.* **20**, 855 (1984).
48. Anderson, B.C., Andrews, G.D., Arthur, P., Jacobson, H.W., Melby, L.R., Playtis, A.J. and Sharkeys, W.H. *Macromolecules* **14**, 1599 (1981).
49. Lutz, P., Masson, P., Beinert, G. and Rempp, P. *Polym. Bull.* **12**, 79 (1984).
50. Hatada, K., Shinozaki, T., Ute, K. and Kitayama, T. *Polym. Bull.* **19**, 231 (1988).
51. Hatada, K., Kitayama, T., Ute, K., Masuda, E., Shinozaki, T. and Masanori, Y. *Polym. Bull.* **21**, 165 (1989).
52. Gnanou, Y. and Rempp, P. *Makromol. Chem.* **189**, 1997 (1988).
53. Waack, R. and Dovan, M.A. *Polymer* **2**, 365 (1961).
54. Waack, R. and Dovan, M.A. *J. Org. Chem.* **32**, 3395 (1967).
54. Hatada, K., Shinozaki, T., Ute, K. and Kitayama, T. *Polym. Bull.* **19**, 231 (1988).
56. Suzuki, T. and Okawa, T. *Polym. Commun.* **29**, 225 (1988).
57. Suzuki, T. *Polymer* **30**, 333 (1989).
58. Maugh, T.H. *Science*, **222**, 259 (1983).
59. Hertler, W.R., Sogah, D.Y., Webster, O.W. and Trost, B.M. *Macromolecules* **17**, 1417 (1984).
60. Webster, O.W., Hertler, W.R., Sogah, D.Y., Farnham, W.B. and Rajanbabu, T.V. *J. Macromol. Sci., Chem.* **A21**, 943 (1984).
61. Sogah, D.Y. and Webster, O.W. *J. Polym. Sci., Polym. Lett. Ed.*, **21**, 927 (1983).
62. Chen, G.M. *Polym. Prepr.* **29**(2), 46 (1988).
63. Asami, R., Kondo, Y. and Takaki, M. *Polym. Prepr.* **27**(1), 186 (1986).
64. Asami, R., Kondo, Y. and Takaki, M. in *Recent Advances in Anionic Polymerization* (Ed. Hogen-Esch, E.T. and Smid, J.) Elsevier, New York, pp. 381–392 (1987).
65. Yamashita, Y. and Chiba, K. *Polym. J.* **4**, 200 (1973).
66. Franta, E., Reibel, L., Lehmann, J. and Penczek, S. *J. Polym. Sci., Polym. Symp.* **56**, 139 (1976).
67. Smith, S. and Hubin, A.J. *J. Macromol. Sci., Chem.* **A7**, 1399 (1973).
68. Smith, S., Schultz, W.J. and Newmark, R.A. *ACS Symp. Ser.* **59**, 13 (1977).

69. Tezuka, Y. and Goethals, E.J. *Eur. Polym. J.* **18**, 991 (1987).
70. D'Haese, F. and Goethals, E.J. *Brit. Polym. J.* **20**, 103 (1988).
71. Kress, H.J., Stix, W. and Heitz, W. *Makromol. Chem.* **185**, 173 (1984).
72. Goethals, E.J. *Makromol. Chem., Macromol. Symp.* **6**, 53 (1986).
73. Brzezinska, K., Szymanski, R., Kubisa, P. and Penczek, S. *Makromol. Chem., Rapid Commun.* **7**, 1 (1986).
74. Penczek, S., Kubisa, P. and Szyzmanski, R. *Makromol. Chem., Macromol. Symp.* **3**, 203 (1986).
75. Kobayashi, S., Masuda, E., Shoda, S. and Shimano, Y. *Macromolecules* **22**, 2878 (1989).
76. Jewel, B.S., Riffle, J.S., Allison, D. and McGrath, J.E. *Polym. Prepr.* **30**(1), 295 (1989).
77. Miyamoto, M., Sawamoto, M. and Higashimura, T. *Macromolecules* **18**, 123 (1985).
78. Sawamoto, M. and Higashimura, T. *Makromol. Chem., Macromol. Symp.* **3**, 83 (1985).
79. Sawamoto, M., Enoki, T. and Higashimura, T. *Macromolecules* **20**, 1 (1987).
80. Kennedy, J.P. and Smith, R.A. *J. Polym. Sci., Polym. Chem. Ed.* **18**, 1523 (1980).
81. Wang, B., Mishra, M.K. and Kennedy, J.P. *Polym. Bull.* **17**, 213 (1987).
82. Faust, R., Nagy, A. and Kennedy, J.P. *J. Macromol. Sci., Chem.* **A24**, 595 (1987).
83. Faust, R., Zsuga, M. and Kennedy, J.P. *Polym. Bull.* **21**, 125 (1989).
84. Nemes, S. and Kennedy, J.P. *Polym. Bull.* **21**, 293 (1989).
85. Kennedy, J.P., Liao, T.P., Guhaniyogi, S. and Chang, V.S.C. *J. Polym. Sci., Polym. Chem. Ed.* **20**, 3219 (1982).
86. Nuyken, O., Pask, S.D., Vischer, A. and Walter, M. *Makromol. Chem., Macromol. Symp.* **3**, 129 (1986).
87. Burgess, E.J., Cunliffe, A.V., Richards, D.H. and Thompson, D. *Polymer* **19**, 334 (1978).
88. Nuyken, O., Pask, S.D., Vischer, A. and Walter, M. *Makromol. Chem.* **186**, 173 (1985).
89. Asami, R., Takaki, M., Kyuda, K. and Asakura, E. *Polym. J.* **15**, 139 (1983).
90. Asami, R., Takaki, M., Kita, K. and Asakura, E. *Polym. Bull.* **2**, 713 (1980).
91. Goethals, E.J. and Vlegels, M. *Polym. Bull.* **4**, 521 (1981).
92. Schulz, R.C. and Schwarzenbach, E. *Makromol. Chem., Macromol. Symp.* **13/14**, 495 (1988).
93. Miyamoto, M., Naka, K., Tokumizu, M. and Saegusa, T. *Macromolecules* **22**, 1604 (1989).
94. Kobayashi, S., Masuda, E., Shoda, S. and Shimano, Y. *Macromolecules* **22**, 2878 (1989).
95. Uryu, T., Yamanaka, M., Date, M., Ogawa, M. and Hatanaka, K. *Macromolecules* **21**, 1916 (1988).
96. Higashimura, T., Aoshima, S. and Sawamoto, M. *Makromol. Chem., Macromol. Symp.* **3**, 99 (1986).
97. Sawamoto, M., Takashi, E. and Higashimura, T. *Polym. Bull.* **16**, 117 (1986).
98. Kennedy, J.P. and Liao, J.P. *Polym. Bull.* **5**, 11 (1981).
99. Farona, M.F. and Kennedy, J.P. *Polym. Bull.* **11**, 359 (1984).
100. Konter, W., Bomer, B., Kohler, K.H. and Heitz, W. *Makromol. Chem.* **182**, 2619 (1981).
101. Guth, W. and Heitz, W. *Makromol. Chem.* **177**, 1835 (1976).
102. Reference 6, pp. 317–320.
103. Ishizu, K., Ono, T., Fukutomi, T. and Shiraki, K. *J. Polym. Sci., Polym. Lett. Ed.* **25**, 131 (1987).
104. Fukutomi, T., Ishizua, K. and Shiraki, K. *J. Polym. Sci., Polym. Lett. Ed.* **25**, 175 (1987).
105. Breitenbach, J.W., Olaj, O.F., Kuchner, K. and Horacek, H. *Makromol. Chem.* **87**, 295 (1965).
106. Haas, H.C., Schuler, N.W. and Kolesinski, H.S. *J. Polym. Sci., Part A-1* **5**, 2964 (1967).
107. British Patent, 957,652 (1964).
108. Guth, W. and Heitz, W. *Makromol. Chem.* **177**, 3159 (1976).
109. Schnecko, H., Degler, G., Dongowski, H., Caspary, R., Angerer, G. and Ng, S. *Angew. Makromol. Chem.* **70**, 9 (1978).
110. Bevington, J.C. and Huckerby, T.N. *Macromolecules* **18**, 176 (1985).
111. Boutevin, B. and Pietrasanto, Y. in *Comprehensive Polymer Science* (Eds. Allen, G. and Bevington, J.C.) Pergamon Press, Oxford, Vol. 3, Ch. 14 (1989).
112. Starks, C.M. *Free Radical Telomerization*, Academic Press, New York, p. 113 (1974).
113. Joyce, R.M., Handford, W.E. and Harmon, J. *J. Amer. Chem. Soc.* **70**, 2529 (1948).
114. Uri, N. *Chem. Rev.* **50**, 375 (1950).
115. Harmon, H., Ford, T.A., Handford, W.A. and Joyce, R.M. *J. Amer. Chem. Soc.* **72**, 2213 (1950).
116. Uraneck, C.A., Hsieh, H.L. and Buck, O.G. *J. Polym. Sci.* **46**, 535 (1960).
117. Otsu, T., Matsunaga, T., Kuriyama, A. and Yoshioka, M. *Eur. Polym. J.* **25**, 643 (1989).

118. Bledzki, A. and Braun, D. *Makromol. Chem.* **182**, 1047 (1981).
119. Bledzki, A., Balard, H. and Braun, D. *Makromol. Chem.* **182**, 1057 (1981).
120. Bledzki, A., Braun, D. and Titzschkau, K. *Makromol. Chem.* **184**, 745 (1983).
121. Otsu, T., Yoshida, M. and Tazaki, T. *Makromol. Chem., Rapid Commun.* **3**, 133 (1982).
122. Bailey, W.J., Chen, P.Y., Chen, S.C., Chiao, W.B., Endo, T., Gapud, B., Kuruganti, Y., Lin, Y.N., Ni, Z., Pan, C.Y., Shaffer, S.E., Sidney, L., Wu, S.R., Yamamoto, N., Yamazaki, N., Yonezawa, K. and Zhou, L.L. *Makromol. Chem., Macromol. Symp.* **6**, 81 (1986).
123. Enikolopyan, N.S., Smirnov, B.R., Ponomarev, G.V. and Belgovskii, I.M. *J. Polym. Sci., Polym. Chem. Ed.* **19**, 879 (1981).
124. Burczyk, A.F., O'Driscoll, K.F. and Rempel, G.L. *J. Polym. Sci., Polym. Chem. Ed.* **22**, 3255 (1984).
125. Caciola, P., Hawthorne, D.G., Laslett, R.L., Rizzardo, E. and Solomon, D.H. *J. Macromol. Sci., Chem.* **A23**, 839 (1986).
126. Harris, F.W., Pamidimukkala, A., Gupta, R., Das, S., Wu, T. and Mock, G. *J. Macromol. Sci., Chem.* **A21**, 1117 (1984).
127. Droske, J.P. and Stille, J.K. *Macromolecules* **17**, 1 (1984).
128. Percec, V. and Auman, B.C. *Makromol. Chem.* **185**, 616 (1984).
129. Viswanathan, R., Johnson, B.C. and McGrath, J.E. *Polymer* **25**, 1827 (1984).
130. Risse, W., Heitz, W., Freitag, D. and Bottenbruch, L. *Makromol. Chem.* **186**, 1835 (1985).
131. Nava, H. and Percec, V. *J. Polym. Sci., Polym. Chem. Ed.* **24**, 965 (1986).
132. Nuyken, O. and Siebzehnruebl, F. *Makromol. Chem.* **189**, 541 (1988).
133. Shchori, E. and McGrath, J.E. *Polym. Prepr.* **20**, 634 (1979).
134. Michel, A., Castenada, E. and Goyot, A. *J. Macromol. Sci., Chem.* **A12**, 227 (1978).
135. Odinokov, V.N., Ignatyuk, V.K., Tolstikov, G.A., Monakov, Y.B., Berg, A.A., Shakirova, A.M., Rafikov, S.R. and Berlin, A.A. *Bull. Acad. Sci., USSR, Div. Chem. Sci.* **25**, 1475 (1976).
136. Keller, A. and Udagawa, Y. *J. Polym. Sci., Part A-2* **10**, 221 (1972).
137. Melby, L.R. *Macromolecules* **11**, 50 (1978).
138. Tanaka, Y., Shimizu, M. and Kageyu, A. Japanese Patent 63/48305 (1988) (C.A. 109(2):7169q).
139. Kapko, J. and Huczkowski, P. *J. Appl. Polym. Sci., Appl. Polym. Symp.* **35**, 67 (1979).
140. Rehner, J. and Gray, P. *Ind. Eng. Chem., Anal. Ed.* **17**, 367 (1945).
141. Galo, S.G., Wiese, H.K. and Nelson, J.F. *Ind. Eng. Chem.* **40**, 1277 (1948).
142. Lee, T.S., Kolthoff, J.M. and Hohnson, E. *Anal. Chem.* **22**, 997 (1950).
143. Ebdon, J.R., Flint, N.J. and Hodge, P. *Eur. Polym. J.* **25**, 759 (1989).
144. Guizard, C. and Cheradame, H. *Eur. Polym. J.* **17**, 121 (1981) and references cited therein.
145. Baily, W.J., Endo, T., Gapud, B., Lin, Y.N., Ni, Z., Pan, C.Y., Schaffer, S.E., Wu, S.R., Yamazaki, N. and Yonezawa, K. *J. Macromol. Sci., Chem.* **A21**, 979 (1984).
146. Bailey, W.J., Chen, P.Y., Chiao, W.B., Endo, T., Sidney, L., Yamamoto, N., Yamazaki, N. and Yonezawa, K. in *Contemporary Topics in Polymer Science* (Ed. Shen, M.), Plenum Press, New York, Vol. 3, p. 29 (1979).
147. Bailey, W.J. *Kobunshi*, **30**, 331 (1981).
148. Macosko, C.W. *RIM, Fundamentals of Reaction Injection Moulding*, Hanser, Munich (1989).
149. Huet, J.M. and Marechal, E. *Eur. Polym. J.* **10**, 757 (1974).
150. Friedman L. and Schechter, H. *Tetrahedron. Lett.* 238 (1961).
151. Kennedy J.P. and Iza, H. *J. Polym. Sci., Polym. Chem. Ed.*, **21**, 1033 (1983).
152. Pourdjavadi, A., Madec, P.J. and Marechal, E. *Eur. Polym. J.* **20**, 311 (1984).
153. Gagnebien, D., Madec, P.J. and Marechal, E. *Eur. Polym. J.* **21**, 301 (1985).
154. Geckeler K. and Bayer, E., *Polym. Bull.* **1**, 691 (1979).
155. Castaldo, L., Maglio, G. and Palumbo, R. *J. Polym. Sci., Polym. Lett. Ed.* **16** 643 (1978).
156. Leriche, Ch., Michaud, Ch. and Marechal, E. *Bull. Soc. Chim.* **1977**, 717.
157. Konter, W., Bomer, B., Kohler, K.H. and Heitz, W. *Makromol. Chem.* **182**, 2619 (1981).
158. Heitz, W., Ball, P. and Lattekamp, M. *Kautsch. Gummi. Kunstst.* **34**, 459 (1981).
159. De Visser, A.C., Gregonis, D.E. and Driessen, A.A., *Makromol. Chem.* **179**, 1855 (1978).
160. Kennedy, J.P. and Hiza, M. *J. Polym. Sci., Polym. Chem. Ed.* **21**, 3573 (1983).
161. Boutevin, B., Piettrasanta, Y., Taha, M. and El Sarraf, T. *Polym. Bull.* **10**, 157 (1983).
162. Plate, N.A., Valuev, L.I. and Chupov, V.V., *Pure Appl. Chem.* **56**, 1351 (1984).
163. Okamoto, Y., Shohi H. and Yuki, H. *J. Polym. Sci., Polym. Lett. Ed.* **21**, 601 (1983).
164. Kennedy, J.P. and Hiza, M., *Polym. Bull.* **10**, 146 (1983).

165. Hudecek, S., Spevacek, J., Hudeckova J. and Mikesowa, J. *Polym. Bull.* **3**, 143 (1980).
166. Chujo, Y., Murai, K., Yamashita, Y. and Okumura, Y. *Makromol. Chem.*, **186**, 1203 (1982).
167. Yamashita, Y., Tsai, H.C. and Tsukhara, Y., *Makromol. Chem. Suppl.* **12**, 51 (1985).
168. Arai, T. and Kawase, S. (Soken Chemical Engineering Co.), European Patent Application 248,574 (1987) (C.A. 108(22): 187481w).
169. Murai, K., Kojima S. and Azuma T. (Toa Gosei Chemical Industry Co.), Japanese Patent 61/232,408 (1987) (C.A. 109(2):7141z).
170. Solomon, D.H., Cacioli, P. and Moad, G. *Pure Appl. Chem.* **57**, 985 (1985).
171. Revillon, A. and Hamaide, T. *Polym. Bull.* **6**, 325 (1982).
172. Nava, H. and Percec, V. *J. Polym. Chem., Polym. Chem. Ed.* **24**, 965 (1986).
173. Percec, V. and Auman, B.C. *Makromol. Chem.* **185**, 1867 (1984).
174. Percec, V., Rinaldi P.L. and Auman, B.C. *Polym. Bull.* **10**, 215 (1983).
175. Greber, G. and Reese, E. *Makromol. Chem.* **55**, 96 (1962).
176. Brosse, J.C., Derouet, D. and Nganie, A., *Eur. Polym. J.* **19**, 55 (1983).
177. Kobayashi, K., Sumitomo, H. and Ina, Y. *Polym. J.* **17**, 567 (1985).
178. Dournel, P., Randrianalimanana, E., Deffieux, A., and Fontanille, M. *Eur. Polym. J.* **24**, 843 (1988).
179. Berlinova, I.V. and Panayotov, I.M., *Makromol. Chem.* **188**, 2141 (1987).
180. Chujo, Y., Kobayashi, H. and Yamashita, Y. *J. Polym. Sci., Polym. Chem. Ed.* **27**, 2007 (1989).
181. Chujo, Y., Kobayashi H. and Yamashita, Y. *Polym. J.* **20**, 407 (1988).
182. Merker, R.L., Scott, M.J. and Haberland, G.G. *J. Polym. Sci.* **2**, 31 (1964).
183. Athey, Jr., R.D. *Prog. Org. Coat.* **7**, 289 (1979).
184. Deleens, G., Foy, P. and Marechal, E. *Eur. Polym. J.* **13**, 353 (1977).
185. McGrath, J.E., Ward, T.C., Shchori, E., Wnuk, A.J., Wiswanathan, R., Riffleand J.S. and Davidson, T.F. *Polym. Prepr.* **19**, 109 (1978).
186. McGrath, J.E., Ward, T.C., Shchori, E. and Wnuk, A.J. *Polym, Eng. Sci.* **17**, 647 (1977).
187. Vaughn, H.A. *J. Polym. Sci., Part B* **8**, 191 (1969).
188. Johnson, R.N., Farnham, A.G., Clendinning, R.A., Hale, W.F. and Meriam, C.N. *J. Polym. Sci., Part A-1* **5**, 2399 (1967).
189. Fradet A. and Marechal, E. *Eur. Polym. J.* **14**, 749 (1978).
190. Huet, J.M. and Marechal, E. *Eur. Polym. J.*, **10**, 757 (1974).
191. Laverty, J.J. and Garlund, Z.R. *J. Appl. Polym. Sci.*, **26**, 3657 (1981).
192. Muhlbach, K. and Percec, V. *J. Polym. Sci., Polym. Chem. Ed.* **25**, 2605 (1987).
193. Cameron, G.G. and Chisholm, M.S. *Polymer* **26**, 437 (1985).
194. Ito, K., Tsuchida, H., Hayashi, A., Kitano, T., Yamada, E. and Matsumoto, T. *Polym. J.* **17**, 827 (1985).
195. Cameron, G.G. and Chisholm, M.S. *Polymer* **27**, 1420 (1986).
196. Ito, K., Tsuchida, H. and Kitano, T. *Polym. Bull.* **15**, 425 (1986).
197. Asami, R. Takaki M. and Matsuse, T. *Makromol. Chem.* **190**, 45 (1989).
198. Asami, R., Takaki M. and Moriyama, Y., *Polym. Bull.* **16**, 125 (1986).
199. Niwa, M., Akahori, M. and Nishizawa, S. *J. Macromol. Sci., Chem.* **A24**, 1423 (1987).
200. Percec, V. and Wang, J.H. *Polym. Prepr.* **29**, 294 (1988).
201. Tsukahara, Y., Mizuno, K., Segawa, A. and Yamashita, Y. *Macromolecules* **22**, 1546 (1989).
202. Stejskal, J., Kratochvil P. and Jenkins, A.D. *Macromolecules* **20**, 181 (1987).
203. Stejskal, J., Starkova, D., Kratochvil, P., Smith, S.D. and McGrath, J.E. *Macromolecules* **22**, 861 (1989).
204. Hamaide, T., Revillon A. and Guyot, A. *Eur. Polym. J.* **23**, 27 (1987).
205. Hamaide, T., Revillon A. and Guyot, A. *Eur. Polym. J.* **23**, 787 (1987).
206. Gnanou Y. and Lutz, P. *Makromol. Chem.* **190**, 577 (1989).
207. DeSimone, J.M., Hellstern, A.M., Ward, T.C., McGrath, J.E., Smith, S.D., Gallagher, M.P., Krukonis, V.J., Stejskal, J., Strakova, D. and Kratochvil, P. *Polym. Prepr.* **29**, 116 (1988).
208. Steinle, E.C., Critchfield, F.E., Castro, J.M. and Macosko, C.W. *J. Appl. Polym. Sci.* **25**, 2327 (1980).
209. Bazuin, C.G. and Eisenberg, A. *J. Polym. Sci., Polym. Phys. Ed.* **24**, 1021 (1986).
210. Portnoy, R.C., Werlein, E.R., Lundberg, R.D. and Peiffer, D.G. *Proc. Amer. Chem. Soc., Div. Polym. Mat. Sci. Eng.* **55**, 860 (1986).
211. *Research and Technology*, ICI Chemicals and Polymers Ltd. (Eds. Gamlen, P.H. and Lane, R.M.), Runcorn, pp. 20–26 (1988).

212. Kawakami, Y., Aoki, T. and Yamashita, Y., *Kobunshi Ronbunshu* **43**, 741 (1986).
213. Rice, D.E. and Ihlenfield, J.V. (Minnesota Mining and Manufacturing Co.), Canadian Patent Application 1,201,247 (1986) (C.A. 106(6):33960y).
214. Cowie, J.M.G., Martin, A.C.S. and Firth, A.M. *Brit. Polym. J.* **20**, 247 (1988).
215. Good, W.R., Mikes, J. and Sikora, J. (Ciba-Geigy, A.-G.), European Patent Application 200,213 (1986) (C.A. 106 (26):21960p).
216. Krampe, S.E., Moore, C.L. and Taylor, C. (Minnesota Mining and Manufacturing Co.), European Patent Application 202,831 (1986) (C.A. 107(4):28425u).
217. Sugo, T., Okamoto, J., Tazaki, S. and Onohara, M. (Japan Atomic Energy Research Institute, Sumitomo Bakelite Co.), Japanese Patent 62/109575 (1987) (C.A. 108(22):192808n).
218. Ceska, G.W. (Pony Industries Inc.), US Patent 4,722,976 (1988) (C.A. 108(24):205966q).
219. Chujo, Y., Hiraiwa, A., Kobayashi, H. and Yamashita, Y. *J. Polym. Sci., Polym. Chem. Ed.* **26**, 2991 (1988).

Index

activated monomer mechanism 13, 133, 173
acylation 153
aldol group transfer polymerization 24, 57, 65, 132
anionic polymerization 10, 23, 67, 165
  comparisons with group transfer polymerization 67
  living 12, 67, 108
  macromers by 167
  of macromers 188
  of methyl methacrylate 23
  telechelics by 165
  transformation to cationic 112
  transformation to metathesis 130
  transformation to radical 120
  transformation to Ziegler–Natta 129
arborols 17
aromatic polyamides 15
aromatic polyesters 15

biosynthetic routes 18
block copolymers 4, 9, 10, 13, 14, 61, 185
  AB 61, 114, 119, 120, 122, 125, 127, 130, 132, 134
  ABA 62, 115, 119, 120, 122, 125, 127, 132, 134
  applications 71
  BAB 61, 119, 120
  by aldol group transfer polymerization 64, 65
  by cationic polymerization 109
  by coupling 64
  by group transfer polymerization 40, 61
  by living anionic polymerization 108
  by metathesis ring-opening 63, 83
  by radical polymerization 4
  by transformation reactions 105–134
  from telechelics 185
bromination 141, 151, 157

carbocationic polymerization *see* cationic polymerization
catalysts
  in aldol group transfer polymerization 58
  in group transfer polymerization 35
  in ring-opening metathesis polymerization 80
cationic polymerization 10, 13, 169
  living 14, 109, 170
  transformation to anionic 116
  transformation to radical 126
chain scission 181
chain transfer *see* transfer agent
chemical modification of polymers 138, 183
  end-groups 183
  general aspects 139
  main chain 140
  pendant groups 145
  polyacenaphthylene 157
  polyacrylates 147
  polyacrylonitrile 148
  polyalkenes 141
  polybutadiene 181
  polydienes 141
  poly(methyl methacrylate) 148
  poly(N-vinyl carbazole) 157
  poly(N-vinyl pyridine) 156
  polysulphones 144
  polystyrene 151
  poly(2-vinyl thiophene) 157
  poly(vinyl alcohol) 145, 146
  poly(vinyl chloride) 143, 181
  reasons for 139
chlorination 141, 143
chloromethylation 151
comb polymers 71
conjugate addition *see* Michael addition
coordination catalysis *see* Ziegler–Natta polymerization

copolymerization
  alternating 9
  anionic 10
  group transfer 59
  radical 8
cyanomethylation 142
cyclization 13, 27, 39, 49, 148
cyclopolymerization 39

daughter polymer 5
dehydrochlorination 143
dendrimers 17

end-groups
  acetylenic 180
  acrylic 173, 184
  amino 165, 177, 184
  anhydride 185
  carboxy 32, 66, 165, 177, 181, 182, 183
  chloroformate 183
  ester 170, 176
  fumarate 185
  halide 66, 165, 177
  hydroxy 32, 66, 165, 168, 171, 176, 181, 183
  imidazole 175
  isocyanate 183
  methacrylic 173
  nitrile 184
  nitro 183
  norbornenyl 174
  phenyl 66
  styryl 169, 174
  thiol 180
  trimethyl siloxane 177
  vinyl ether 173
'ene' reaction 142
enolate 26, 28, 33, 35
esterification 146

Friedel–Craft reactions 151, 153

graft copolymers 185, 187
Grignard reagent 113, 165, 166
group transfer polymerization (GTP) 12, 22, 168
  acrylates 55
  applications 70
  backbiting 40
  batch polymerizations 51
  catalysts 35
  chain transfer 50
  comparisons with anionic polymerization 25, 52, 67
  degree of polymerization 29, 50
  general features 28
  initiation 46

initiators 31
isomerization 40
kinetic data 46, 48, 49
kinetics 43
macromers by 169
mechanism 29, 43
methacrylates 43
molecular weight 30
monomers 36
nature of propagating species 43
probability of propagation 41
propagation 46
reaction conditions 30
reactions in acetonitrile 54
telechelics by 66, 168
temperature effects 30
termination 29, 31, 39, 40, 48
terminology 24
transformation to radical 132

hydroboration 142
hydroformylation 143
hydrolysis 148, 150, 181, 182

'immortal' polymers 13
iniferters 3, 177
ionomers 190

ladder polymer 71, 148
liquid–crystalline polymers 16, 37
lithiation 143, 145, 153, 157
livingness enhancement 36, 56
living polymers/polymerization 3, 12, 14, 22, 64, 67, 83, 108, 170

macroinitiator 4, 63, 65, 133
macromers 163
  anionic polymerization of 188
  by anionic polymerization 167
  by cationic polymerization 172
  by chain scission 182
  by end-group modifications 183
  by group transfer polymerization 169
  by radical polymerization 178
  by stepwise polymerization 181
  definition 163
  group transfer polymerization of 188
  radical polymerization of 187
  reaction injection moulding of 190
  reactivity 185
  steric shielding 187, 188
  synthesis 167, 169, 172, 178, 181, 183
  uses of 190
macromonomer see macromer
mercuriation 155, 157
metallation 143, 151, 153, 155
Michael addition 18, 25, 163

INDEX

neighbouring group effects 140
nitro-mercuriation 142

oligomers *see* telechelic oligomers and macromers
oxidation 146, 181, 183
ozonolysis 181

PEEK 17
photo-Fries rearrangement 15
polyarylate 15
poly(β-hydroxybutyrate) 19
polyketone 16
polymer-supported reagents 150, 156
polyphenylene 19
polysulphone 16

radiation grafting 141
radical polymerization
  general features 2
  mechanism 2
  ring-opening 6, 182
  synthesis of macromers by 178
  synthesis of telechelics by 174
  transformation to active monomer mechanism 133
  transformation to ionic 126
random copolymers 19, 60
reaction injection moulding (RIM) 10, 189
reactivity ratios 8, 187
ring-opening metathesis polymerization 76
  activators 81
  bias (B) of polymer 100
  bicyclo-2.2.1-heptene 80
  catalysts 80
  cross-metathesis 88, 89
  cyclooctene 95
  cyclopentene 76, 77, 95
  Fischer and Casey carbenes 82, 90, 93
  funtionalized monomers 79
  mechanism 84
  metallocyclobutanes 91
  molecular weight distribution 94
  monomers 77
  non-pairwise mechanisms 87
  norbornene 97
  pairwise mechanisms 86
  ring-chain equilibrium 96
  scope 76
  secondary 78
  stereochemistry 97, 98
  substituted norbornenes 99
  Tebbe's reagent 83, 92
  transalkylation 84
  transalkylidenation 84
  transformation to aldol group transfer 63, 132
  transition metal carbenes 92

ring-opening polymerization
  anionic 12, 67
  cationic 13
  ceiling temperature 105
  group transfer 39
  ionic and coordination 11
  radical 6, 182
  thermodynamics 104
ring-strain 6, 78, 105

self-replication 5
silyl ketene acetals 27
starburst polymers 17
star polymers 71
stepwise polymerization 14, 178
  macromers by 181
  molecular weight distribution 179
  of telechelic oligomers 186
  telechelics by 179
sulphonation 144

taxogen 177
telechelic oligomers 66, 162
  acrylates and methacrylates 168
  block copolymers from 185
  butadiene 175, 176, 181
  by anionic polymerization 165
  by cationic polymerization 169
  by chain scission 181
  by end-group modification 183
  by group transfer polymerization 66, 168
  by stepwise polymerization 179
  definition 162
  ethylene 176, 177, 181
  lactones 167
  methyl acrylate 175
  methyl methacrylate 169, 175
  olefins 171
  reaction injection moulding of 189
  reactivity of 185
  stepwise polymerization of 185
  styrene 176, 177, 184
  synthesis 164, 168, 169, 175, 179, 181, 183
  tetrahydrofuran 170
  uses 188
  vinyl acetate 175, 176
  vinyl ethers 171
telechelics *see* telechelic oligomers
telogen 177
telomer 177
template 5, 158
template polymerization 5
terminally reactive oligomers *see* telechelic oligomers and macromers
termination
  anionic polymerization 23
  group transfer polymerization 31, 39
  radical polymerization 2, 175, 177

thalliation 155, 157
transalkylation 84
transalkylidenation 84
transfer agent 3, 50, 174
transformation reactions 107
  anionic to cationic 112
  anionic to metathesis 130
  anionic to radical 120
  anionic to Ziegler–Natta 129
  anion to cation coupling 118
  cationic to anionic 116
  cationic to radical 126
  efficiencies 124
  group transfer to radical 132
  historical development 110
  metathesis to aldol group transfer 63, 132
  radical to active monomer 133
  radical to ionic 128
  radical to polypeptide 134
  Ziegler–Natta to radical 131

Wittig reaction 154
Wurtz coupling 113, 143

Ziegler–Natta polymerization 10, 81, 88, 129, 131